The Epigenome
Molecular Hide and Seek

Edited by
S. Beck and A. Olek

The Epigenome

Molecular Hide and Seek

Edited by
S. Beck and A. Olek

WILEY-VCH

WILEY-VCH GmbH & Co. KGaA

Editors:

Dr. Stephan Beck
The Sanger Centre
Wellcome Trust Genome Campus
Hinxton, Cambridge CB 10 1SA
UK

Dr. Alexander Beck
Epigenomics AG
Kastanienallee 24
10435 Berlin
Germany

Library of Congress Card No.: applied for
A catalogue record for this book is available from the British Library.

Bibliographic information published by Die Deutsche Bibliothek
Die Deutsche Bibliothek lists this publication in the Deutsche Nationalbibliografie; detailed bibliographic data is available in the Internet at <http://dnb.ddb.de>.

Printed in the Federal Republic of Germany.
Printed on acid-free paper.

Composition   ProSatz Unger, Weinheim
Printing   betz-druck GmbH, Darmstadt
Bookbinding   Großbuchbinderei J. Schäffer GmbH & Co. KG, Grünstadt

ISBN   3-527-30494-0

Preface

The Human Genome Project (HGP) delivered the location and sequence of all genes – the Human Epigenome Project (HEP) will help to uncover their temporal and spatial expression. By systematically analyzing the Epigenome for (cytosine) methylation patterns, epigenetic markers will be discovered that have the power to distinguish active from non-active genes and healthy from non-healthy tissues. This new knowledge will provide the missing link between genetics, disease and the environment and will be invaluable for the understanding of how our genome functions and for future health care.

Written in popular science style, this book is aimed at anyone interested in the science arising from the human genome project and its implications for future biomedical research. In nine chapters, world-leading scientists discuss the history, current facts and future potential of this most exciting area in genome science at the interplay of nature and nurture. The topics covered stretch from the discovery of epigenetic signals to their use in molecular diagnostics and therapy but also explore their effects or dependency on development, gene regulation, disease, diet and aging. In addition to the described science, the central messages of all chapters have been captured in artistic illustrations that may well serve as future icons for these fast evolving fields of research.

Following the announcement of the human genome draft sequence in June 2000, a "spoof" editorial in a leading science magazine *(Nature Biotechnology)* suggested that a trilogy be required to complete this historic project: [I] "The Draft", [II] "The Closure" and [III] "The Epigenome Strikes Back". With the "Draft" delivered, the "Closure" on track for Spring 2003, the time has come to explore the next frontier, the "Epigenome".

Stephan Beck & Alexander Olek
January 2003

Contents

1
Five Not Four:
History and Significance of the Fifth Base

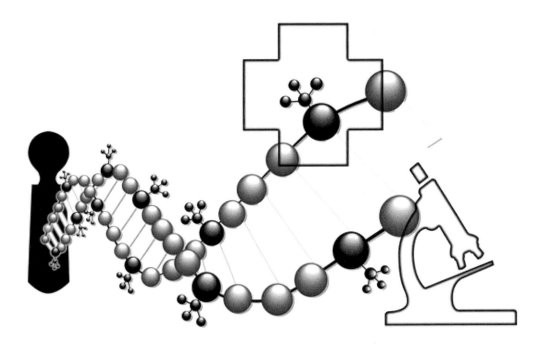

The Key...   to healthcare...   ...and research

1

Five Not Four: History and Significance of the Fifth Base

Douglas S Millar, Robin Holliday, and Geoffrey W Grigg

Summary

The past 30 years have seen a tremendous surge of interest in the field of epigenomics. This work has led to exciting discoveries: new enzymes that can methylate mammalian DNA in both a maintenance and a de-novo manner, the bisulfite modification technique for the direct detection of methyl cytosine residues in genomic DNA, the role of abnormal methylation in disease and the aging process, and, more recently, the link between chromatin remodeling and methylation in the transcriptional repression of gene function. Over the next 30 years, epigenomics will be one of the boom fields of biotechnology with pharmaceutical companies waking up to the fact that the presence or absence of the 5th base has major implications in human health care and the treatment of an ever-increasing aged population.

One of the first major hurdles is the sequencing of the "methylome" or epigenome, which is unlikely to be achieved in the same manner as the sequencing of the human genome. This is because each individual tissue type in the human body has a unique methylation signature, and at least 180 cell types make up the human body. With advances in robotics and automation, however, this task does not seem quite as impossible as it would have been before the race to sequence the human genome began.

Sequencing the human epigenome aside, many questions have at best been only touched upon and many more are still unanswered. How are basic developmental pathways involving methylation established and later modified in a coordinated fashion? Can methylation directly cause gene inactivation or is the process a secondary phenomenon to stably repress gene function? How accurately is the original methylation pattern repaired in a region of DNA damaged by radiation, chemical reactions, or other environmental stimuli? In the next 30 years can the methylation machinery or methylation fingerprints be altered to therapeutically reverse the disease process or indeed the natural process of aging? These questions and many more need to be fully answered before we have true insight into the full impact that epigenomics has on basic human development and disease.

1.1
Historical Introduction

The modified base 5-methyl cytosine (5-mC) was first detected in DNA by Hotchkiss [1]. In subsequent years, many other modified bases were discovered in bacteria, and it became clear that they were an integral part of the modification-restriction systems that are widely distributed in prokaryotes. It was shown that restriction enzymes could distinguish between modified and unmodified DNA substrates.

In plants, approximately 80% of cytosines at CpG doublets are methylated; however, methylation also frequently occurs in CpNpG triplets [2]. Recent studies have also found that methylation of cytosines can occur in non-symmetrical sites such as CpTpA [3]. Despite the overall high global levels of 5-mC, plants contain regions of unmethylated DNA that resemble the *HpaII* tiny fragments characteristic of mammalian genomes. These areas are typically unmethylated in a wide variety of tissues, whether or not the associated gene is expressed [4].

In mammalian DNA, the major modified base is 5-mC, at a level of 2%–5% of all cytosines. Small amounts of other modified bases should not be ruled out, and there is one reliable report that 6-methyl adenine exists in mouse DNA [5]. 5-mC occurs predominantly, but not exclusively, in CpG doublets [6]. In the total mammalian genome, the CpG doublet is very significantly less than what would be expected from the overall base composition of the DNA. However, there are CpG islands associated with structural genes, where such depletion is not observed.

Riggs [7] and Holliday and Pugh [8] proposed that 5-mC in mammalian DNA might have an important role in the regulation of gene expression. They argued that, since restriction enzymes can distinguish between modified and unmodified sites, there might also be regulatory proteins that could make the same distinction at specific DNA sequences in promoter regions. Instead of cutting the DNA, they would recognize the specific methylation signal and would thereby control the presence or absence of transcription at an adjacent gene. These authors also proposed that the pattern of DNA methylation could be inherited, if there was a maintenance DNA methyltransferase that recognized hemimethylated DNA just after replication and methylated the new strand. The same enzyme would not act on nonmethylated CpG doublets. This provided a basis for the epigenetic inheritance of a given pattern of DNA methylation and therefore also of the specific controls of gene expression in given cell types. It could also account for those cases, such as X-chromosome inactivation in female eutherian mammals, where only one of two homologous genes in a diploid cell is active, while the other is inactive. It was already known that these states of activity are very stable in dividing and nondividing cells. In 1975, when these papers appeared, there was no direct evidence for the hypotheses proposed, and the authors therefore did not suggest that methylation, or lack of methylation, was associated with gene activity. However, it later became clear that methylation is associated with the inactivity of genes in almost all cases. For example, in the inactive X chromosome, CpG islands adjacent to genes are methylated, and these same islands are unmethylated on the active X chromosome. It also became clear that the pattern of DNA methylation could be stably inherited, so DNA methyl-maintenance enzymes must exist in cells. However, DNA methyltransferases that were studied in vitro had some activity on nonmethylated substrates in vitro.

In most early experiments, methyl-sensitive restriction enzymes were used, together with an isoschizomer that recognized the same short sequence and cut it whether or not it was methylated. For example, the enzyme HpaII cuts the substrate CCGG, but not if the central CpG is methylated, whereas MspI cuts whether or not methylation is present. It is evident that in all such studies, only a proportion of all 5-mC bases are detected. In the example just cited this is less than 10% of all CpGs. Many published studies in which this method was used have concluded that lack of gene expression in the system studied had nothing to do with DNA methylation. However, it should have been obvious that a full sequence analysis of all methylated and nonmethylated cytosines was necessary before such a conclusion could be reached.

1.2
Sequencing 5-methylcytosine (5-mC) Residues in Genomic DNA

Historically, the first of the chemical methods for sequencing 5-mC arose as a by-product of Maxam and Gilbert's method for sequencing DNA [9]. This method relies on specific but partial cleavage by various reagents at particular base sequences. Hydrazine, one of the reagents used, reacts with cytosine but not with 5-mC, thus allowing differentiation between 5-mC and C sites. Its main drawback was that it required the use of purified radiolabeled restriction fragments of the region of interest.

This protocol was improved by adding to the chemically cleaved fragments an indirect end-labeling procedure [10]. This method, however, still required a relatively large amount of genomic DNA and was somewhat complicated to perform.

The Saluz and Jost procedure [11], which introduced a PCR amplification step and simplified some of the procedures, was a further significant improvement, but it too required 25–50 µg of genomic DNA to produce a signal from a single copy of a eukaryotic gene. This drawback was answered by the ligation-mediated polymerase chain reaction [12]. All these methods, however, recognize an individual cytosine residue by a negative observation – a gap in the sequencing ladder – which is possible to miss when reading the sequence.

The Bisulfite Method
The need for positive and sensitive detection of methylated cytosine residues in genomic DNA was addressed by the development of the bisulfite genomic sequencing protocol. This powerful method depended on the reaction of bisulfite with cytosines in single-stranded DNA which, in an aqueous milieu, are converted to uracil; whereas 5-methylcytosines (5-mC) are unreactive [13]. The modified DNA strands, which now had very few C residues, could be amplified by use of PCR and then either sequenced or, if single half-strand data were required, cloned before sequencing. Direct sequencing without a prior cloning step allowed the average methylation status of individual CpG sites to be obtained, whereas sequencing cloned DNA allowed the analysis of CpG sites on individual half-strand DNA molecules.

The bisulfite method was invented in 1988–1989 [14] and developed into a practical method by Frommer and associates [15]. Several significant improvements to the original method were described subsequently, which greatly improved the sensitivity and removed aberrations caused by secondary structure formation in single-stranded DNA and also demonstrated that the method was applicable even with very small numbers of cells in developmental studies [16–20]. Another study demonstrated that bias may be introduced by the amplification reaction, depending on the sequence of the DNA region of interest [21]; thus, precautions should be taken to prevent false-positive and -negative results.

The principal advantages of the bisulfite method over others are:

- It can provide good DNA sequence information from only a few cells.
- It is a completely general method and provides a positive readout for 5-mCs in the sequencing gel. Since all the C residues are missing from the template but all the mC residues remain, the positions of the latter appear as distinct bands, making them easy to score.
- Single DNA half-strand data on the 5-mC distribution can be determined readily if needed; thus, information about possible hemi-methylated sites or the methylation status of individual alleles useful in analyzing silencing of genes involved in imprinting, normal embryonic development, and cancer can be determined.

The bisulfite genomic sequencing protocol and modifications thereof have revolutionized the field of methylation analysis, yet confusion still comes from the assumption that several authors have made that *all* 5-mC bases are in some way related to gene expression. From the outset, however, this seemed unlikely, and it seemed probable that only a small subset might have a direct regulatory role, with the rest having some other function. One such function appears to be the inactivation of foreign DNA, such as mobile genetic elements, which become inserted into the genome. It is also likely that genes in transgenic animals that become inactivated by a so-called "position effect", are in fact inactivated by DNA methylation. Another function of methylation may be to suppress recombination between repeated genetic sequences in the genome [22]. Without such suppression, many chromosome abnormalities would be generated by crossing-over at meiosis and also during mitosis.

Many authors have argued that DNA methylation may not be important in development, because it was believed that some complex eukaryotes, such as *Drosophila*, have no 5-mC in their DNA. Recently, this belief was shown to be mistaken, because *Drosophila* DNA does contain 5-mC, particularly in the larval stage of development [23, 24]. Indeed, this may strengthen the case for thinking that only a minor fraction of all methylation in mammalian DNA has an important role in development.

1.3
Gene Silencing

It is frequently stated in the literature that DNA methylation is correlated with the inactivation of genes and is sometimes stated that the role of methylation is to "lock in" an inactive state, as a consequence of the silencing of a gene by some other mechanism. However, whatever the mechanism, this is an epigenetic change, spontaneous or induced, that has a specific phenotypic effect; so the term "epimutation" was coined to explain it. When mutations were first studied by geneticists, they had no knowledge of the chemical nature of the gene, let alone how genes produced a particular phenotype. With appropriate genetic analysis, they could deduce that a gene mutation caused a specific phenotype; it was not just correlated with that phenotype. The same applies to methylation changes and gene inactivation. Where a proper analysis can be done, it becomes clear that the methylation of a particular gene causes a particular phenotype, and demethylating agents reactivate that gene. Such analysis has been done in cultured mammalian cells [25], in control elements in maize [26], and in the fungus *Ascobolus* [27]. A study of the structural gene for adenine phospho ribosyl transferase (APRT) in CHO cells demonstrated that two types of inheritance exist: one is due to standard mutations, and the other is due to epimutations [28].

One method of silencing genes by epimutations is to use 5-methyl dCTP after permeabilization of the cells by electroporation [29, 30]. In the experiments with APRT, the epimutations were analyzed by the bisulfite procedure of genomic sequencing [28]. Usually, a single epimutation resulted in extensive methylation of the 16 CpG doublets in the region that was sequenced (within part of the CpG island spanning the promoter region). It is extremely unlikely that such extensive methylation would occur following 5-methyl dCTP treatment. What seems much more plausible is that a low initial level of methylation may be a sufficient trigger for further extensive methylation in the same region of DNA. A similar two-step mechanism may also occur in tumor-suppressor genes (see Sect. 1.5). Further analysis is needed to determine whether the initial epimutation is sufficient to produce the observed phenotype or whether the process of secondary methylation is also necessary. Much more information is also needed in the examples of genes without associated CpG islands. It is possible that a low level of methylation in the promoter region is sufficient to prevent transcription. Furthermore, there may be critical sites where 5-mC blocks gene expression [31, 32]. Some transcription factors are methyl sensitive, which strongly suggests that such sites exist.

It has often been stated that the CpG methylation observed in the silenced genes on the inactive X chromosome is a secondary phenomenon that stably "locks in" the inactive state of the chromosome. It is nevertheless possible that the initial epigenetic switch to inactivation is due to a low level of methylation, perhaps spreading from the inactivation center. What is not in doubt is the effect of another epimutagen, 5-azacytidine [33], which has been shown to reactivate silent genes in a variety of contexts. The evidence suggests that the analog is incorporated into DNA (in its deoxy form), and that the maintenance methyltransferase binds covalently to it [34].

Thus maintenance fails, and even a heavily methylated gene can be fully reactivated in a single step.

DNA methylation is also associated with the silencing of transgenes. Transgenes have been used extensively in the plant industry to improve plants of agronomic significance. However, a major limitation to this technology has been the fact that transgenes are frequently inactivated [35]. One mechanism whereby transgenes are inactivated is promoter methylation [36]; but such silenced transgenes can be subsequently reactivated by 5-azacytidine treatment or by targeted disruption of the plant *ddm1* methyltransferase [37]. Silencing is more common when multiple copies of the gene are inserted, possibly due to pairing of transgenes at the same or different loci [36]. Inverted transgene repeats may also form structures that are recognized more efficiently by the methytransferase enzymes [38]. Other mechanisms to account for the silencing process include RNA-mediated de-novo methylation and differences in the base composition of the transgene and the target organism [37].

1.4
Development

Although it is now generally accepted that the program for development from fertilized egg to adult is under epigenetic control, developmental biologists, by and large, have not concentrated their efforts on the study of methylation changes in particular genes. The methylation model has those features that are necessary for normal development, namely, the switching of gene activities and the imposition of given patterns of DNA methylation in different cell types. The various cell types may be fully differentiated cells or stem cells, which divide to produce both cells that will become differentiated and also more stem cells. It had also been proposed that developmental clocks, controlled by DNA methylation, may be an important component of the developmental program [8, 39, see Sect. 1.7]. These clocks provide a counting mechanism that determines the number of cell divisions that will occur before a subsequent developmental event. For example, the size of the retina is believed to be determined by one such mechanism [40].

In spite of the relative dearth of direct information about a primary role for DNA methylation in unfolding the developmental program, there are many different indications that methylation changes are heavily involved. One of the best-studied areas is genomic imprinting, where specific methylation differences have been shown between male and female gametes (reviewed by Reik, Chap. 4). The DNA sequences in sperm and egg are not by themselves sufficient for normal development of the fertilized egg. Additional epigenetic information is superimposed on the DNA of both sperm and oocyte or egg, and this information is different in the male and female gametes. Each complements the other to produce a normal embryo. This means that for some genes there is one active and one inactive allele in the fertilized egg and early embryo. (A situation that is analogous to the X chromosome dosage compensation mechanisms in females). It is possible that a single copy of a gene important in early development, rather than two copies, facilitates precise switching of gene activities in early cell lineages [41]. If this is so, it is possible that imprinting signals are lost or change later in development. This may be the reason why it is difficult to transfer a somatic cell nucleus to an anucleate egg and then obtain normal cloned offspring. The success rate in this type of nuclear transfer experiment is very low, and when the cloned animals do develop, they often have abnormalities (see Sect. 1.6).

Early studies showed massive changes in genomic methylation during gametogenesis and embryogenesis [42, 43]. It is not known whether these changes affect specific genes or whether they are more global and nonspecific. They are presumably part of the overall reprogramming that must occur in each generation, but a great deal of more detailed analysis will be necessary to reveal the details of such reprogramming. Another compelling reason for believing that DNA methylation is essential for development is that knockout mice lacking a particular DNA methyltransferase enzyme are embryonic lethal [44, 45]. In some ingenious experiments, Donoghue et al. [46] studied the methylation of a transgene in the somites of a mouse embryo. They discovered that a gradient of methylation from one end of the somites to the other exists. Moreover, when cells from each somite were grown in culture, the level of methylation was maintained. This demonstrates an important connection between the actual position of differentiating cells in a developing embryo and their level of methylation.

Several experiments have demonstrated the importance of specific demethylation and the subsequent expression of genes. In cultured mouse cells (10T1/2 and 3T3), the demethylation of the master regulating *Myo D1* gene by azacytidine results in the formation of differentiated adipocytes, chondrocytes, and myotubes [47, 48]. Also, when a methylated gene is tranfected into a cell that normally expresses that gene, both specific demethylation and expression occur. In contrast, when the same methylated gene is transfected into a cell type that does not normally express the gene, then it remains methylated and inactive [49, 50]. A more specific example of demethylation was revealed by genomic sequencing. Estradiol induces the egg protein vitellogenin in the liver of egg-laying hens, which is associated with the specific demethylation of CpG doublets in the promoter region of the estradiol response element (*ERE*) [51]. This raises the important issue of specific demethylation in developmental contexts in response to growth factors, hormones, or other inducing signals. Epigenetic changes induced in this way could occur in groups of receptive cells and could subsequently be inherited through mitotic division.

The study of the effect of environmental signals in the transition from vegetative growth to flowering in plants has shed some light on how such signals direct development by altering methylation patterns. Vernalization is the process in which exposure of a germinating seed or juvenile plant to prolonged periods of low temperature promote flowering in an adult plant [52]. Burns et al. proposed that vernalization is mediated by demethylation in the promoter of gene(s) critical for flowering, since treatment with 5-azacytidine or antisense methyltransferase constructs both promote flowering in the absence of cold treatment [52].

Recently, a key gene in the process, *FLOWERING LOCUS C* (*FLC*), has been identified. *FLC* acts as a repressor of flowering, and results show that levels of *FLC* correlate with the time to flowering. *FLC* mRNA and protein are downregulated by exposure of germinating seeds to low temperatures, the extent of downregulation being proportional to the duration of cold treatment, as is the promotion of flowering. Consistent with the idea that vernalization is regulated via methylation signals, plants containing antisense methyltransferase constructs flower early and have a reduced level of *FLC* transcript. Thus, the downregulation of *FLC* suggests that methylation may block expression of a repressor of *FLC* or perhaps the binding of a repressor to the *FLC* promoter [53].

If we assume that the pattern of DNA methylation changes in a controlled, orderly way throughout development, then it would not be surprising that fully differentiated cells of any one type would have an invariant pattern. This is precisely what has been observed [54, 55; Doerfler, see Chap. 2]. The aim of the human epigenome project (HEP) is to uncover the fine detail of DNA methylation in different cells and tissues. What is also needed is an analysis of the dynamic changes in DNA methylation during gametogenesis, early embryogenesis, and subsequent development. We believe that this will illuminate in many ways the innumerable interactions between, and changes within, cells and tissues, all of which are an integral part of the developmental process.

1.5
Abnormal DNA Methylation in Cancer Cells

It was first proposed nearly a quarter of a century ago that DNA methylation may be an important contributor to the process of carcinogenesis [56]. The first direct evidence for an alteration in methylation levels in cancer was the observation that global methylation levels in cancer were lower than the levels in normal cells [57]. Fienberg and Volgelstein further demonstrated that the decrease in global methylation levels was associated with the activation of cellular oncogenes such as *K-ras* in lung and colon tumors [58]. The hypomethylation in tumor cells most commonly occurs in repetitive and parasitic DNAs, which are characteristically hypermethylated in normal cells. This phenomenon could also correlate with the genetic instability seen in cancer due to the activation of endogenous transposable elements and other mobile elements found in the human genome.

In addition to hypomethylation, cancer cells exhibit regional hypermethylation. Frequent targets of hypermethylation are promoter regions within CpG islands. In 1986 the gene for calcitonin was the first to be identified as being abnormally methylated in cancer cells but not in normal tissues [59]. To date, at least 60 genes have been shown to be abnormally hypermethylated in cancers. A selection of these genes are listed in Table 1.1. This table is by no means exhaustive, since new genes are being identified with an ever-increasing frequency.

Several studies have demonstrated a direct role of abnormal CpG-island promoter methylation in gene silencing in cancer. Gene silencing had been shown in *Von-Hippel Landau* in renal cancer [60], in retinoblastoma [61], for GSTP1 in prostate cancer [62], as well as many in other cancer types as the primary inactivation event, since

Tab. 1.1 A selection of genes found to be hypermethylated in cancer (see http://www3.mdander-son.org/leukemia/methylation/cgi.html for updates to this list)

Gene	Location	Cancer	Comments
14–3-3 Sigma	1p	Breast/gastric	
ABL1	9q34.1	CML/AML	Only methylated when part of bcr-abl translocation
ABO	9q34	Cell lines	
APC	5q21	Colon, gastric, and esophageal	Invasion tumor architecture
Androgen Receptor	Xq11–12	Prostate/colon	Growth factor
BLT1		Various cell lines	
BRAC1	17q21	Breast/ovarian	DNA damage repair
Calcitonin	11p15	Colon, lung. and hematopic	One of first genes hypermethylated in cancer
Caspase 8	2q33–34	Neuroblastoma	Apotosis inhibitor
Caveolin 1	7q31.1	Breast cell lines	
CD44		Prostate	
CFTR	7q31.2	Cell lines	Not primary tumors
COX2	1q25.2–25.3	Colon, breast. and prostate cell lines	Correlates with expression
CSPG2	5q12–14	Colon	Regulated by RB
CX26	13q11–12	Breast cell lines	
Cyclin A1	13q12.3–13	Cell lines	
DBCCR1	9q32–33	Bladder	Also low level in normal bladder
E-cadherin	16q22.1	Breast, gastric, thyroid, SCC, leukemias. and liver	Cell–cell adhesion
Endothelin Receptor B	13q22	Prostate	Growth factor response
EPHA3	3p11.2	Leukemias	
Estrogen Receptor	6q25.1	Colon, liver, heart, lung. and leukemias	Growth factor response
FHIT	3p14.2	Esophageal	
Glypican 3	Xq26	Mesothelioma/ovarian	
GSTP1	11q13	Prostate, liver, colon, breast, and kidney	DNA damage repair
H19	11q15.5	Wilm's	Imprinted gene
H-cadherin	16q24.1–2	Lung/ovarian	
HIC1	17p13.3	Prostate, breast, brain, lung, and kidney	Candidate tumor suppressor gene

Tab. 1.1 (continued)

Gene	Location	Cancer	Comments
hMLH1	2p22	Colon, endrometrial, and gastric	DNA mismatch repair
HOXA5	7p14.2–15	Breast	
IGF2	11p15.5	Colon/AML	Imprinted gene
IGFBP7	4q12	SV40-induced hepatocarcinoma	Normal and primary tumors?
IRF7		Cell lines	
LKB1	19p13.3	Colon, testicular, and breast	Serine/threonine protein kinase
MDGI	1p33–35	Breast	
MDR1	7q21.1	Drug-sensitive leukemias	Primary tumors?
O^6MGMT	10q26	Brain, colon, lung, and breast	DNA damage repair
MUC2	11p15.5	Colon cell lines	Primary tumors?
MYOD1	11p15.4	Colon, breast, bladder, and lung	One of first cancer related found
N33	8p22	Colon, brain, and prostate	Oligosaccharyl transferase
NEP	3q21–27	Prostate	
NIS	19p13.2–12	Thyroid cell lines	Not primary tumors
P14/ARF	9q21	Colon cell lines	Cell cycle control
P15	9q21	AML/ALL	Cell cycle control
P16	9q21	Lung, colon, lymphoma, bladder, and many others	Cell cycle control
P57/KIP2	11p15.5	Gastric cell lines	Cell cycle control
PAX	11p13	Colon	
PgR	11q22	Breast	Effect on transcription?
RASSF1	3p21.3	Lung	Growth factor response
RB1	13q14	Retinoblastoma	Cell cycle control
TESTIN	7q31.2	Hematopoetic	
TGFBR1	9q33–34	Gastric	
TIMP3	22q12.1–13.2	Brain/kidney	
VHL	3p25	Renal/common in solid and liquid tumors	Angiogenesis stimulator
WT1	11p13	Breast, colon, and Wilm's	Correlation with expression?

treatment of cell lines with the drug 5-azacytidine (which inhibits DNA methyltrans-ferases) results in reactivation of the silenced genes [60]. Concurrent methylation of multiple genes has also been observed in human cancers, and evidence suggests that certain subsets of genes may become methylated in a cancer-specific fashion [63]. Moreover, methylation inactivation of gene expression seems to be a very early event in tumorogenesis, since methylation has also been demonstrated in preneo-plastic lesions in colon cancer, with a correlation between methylation and disease progression.

Most silenced tumor suppressor genes in cancer cells have heavy methylation in CpG-island regions. This methylation might be a two-step process, as is likely for the APRT gene in CHO cells [28]. An initial low level of de-novo methylation would be followed by the imposition or spreading of additional methylation, which switches the gene into a stable inactive state. Experimental evidence suggests that the initial methylation may be due to the reincorporation of 5-methyl dCMP, which is pro-duced in the cell following the repair of spontaneous lesions in DNA [64]. Normally, 5-methyl dCMP is deaminated to TMP by a specific enzyme. However, many en-zymes have altered activity in cancer cells [65], so it is possible that some 5-methyl dCMP is re-incorporated into DNA as the di- and tri-phosphonucleotides. The main evidence for this came from the isolation of a strain of CHO cells with a significantly reduced level of 5-methyl dCMP deaminase. This strain also had very high frequency of spontaneous epimutations at the two loci that were tested. Radiolabelling with tri-tiated methyl deoxycytidine demonstrated that it was incorporated into DNA in this epimutator strain [64].

1.6
Nuclear Transfer

In all mammalian species when cloning has been successful, whether using embryo-nic cells or somatic cells as the donor, at best only a few percent of the cloned em-bryos survived till birth, and even then, many died soon after birth [66]. Embryos may also suffer from a variety of genetic abnormalities, including respiratory distress and circulatory problems [67], and may be more susceptible to cancers [68]. Even the "healthy" survivors may suffer a wide range of conditions, including immune dys-function or kidney or brain abnormalities, which may contribute to death at a later stage of development [69, 70]. In addition, a recent report showed that mice pro-duced by nuclear transfer technology exhibited shorter life spans than IVF controls [68]. Such abnormalities have also been seen in humans and in mice, where targeted disruption of imprinted genes has occurred [71]. This suggests that the high failure rate using nuclear-transfer technology may be due to incomplete epigenetic repro-gramming of the embryo [67].

Three recent studies by Kang looked at the methylation profiles in cloned bovine and pig embryos and compared the methylation with that of normal (IVF-derived) embryos. Demethylation was observed in the Bov-B line repetitive elements in IVF bovine embryos, but not in nuclear-transfer cloned embryos, which showed similar patterns to the donor cells [72]. A follow-up study found that the inefficient demethylation observed in the nuclear-transfer embryos could be reprogrammed by the presence of oocytic nuclei [73]. These results indicate that some factors provided by the oocytic nuclei may assist the demethylation of satellite sequences in normal development. Supporting evidence for this comes from a recent study in which Tada et al. [74] fused adult thymocytes with embryonic stem cells (ES). The inactive X chromosome derived from the female thymocyte adopted some characteristics of the active X chromosome [74]. In addition *Oct 4*, which is normally silenced in the adult thymocyte, was reactivated after fusion. Interestingly, the somatic DNA methylation pattern of the imprinted genes *H19* and *Igf2r* was maintained in the hybrids but erased in hybrids between ES cells and embryonic germ cells [74].

When the epigenetic reprogramming in pig embryos generated by nuclear transfer and IVF technology was analyzed, it was found that, unlike in bovine embryos, the methylation patterns of the centromeric satellite and Pre-1 SINE elements were similar in the nuclear-transfer and IVF-derived embryos. These results could indicate that species-specific differences in modifying the epigenetic status of cloned donor genomes may exist [75].

In addition, other studies have shown that the epigenetic information established during gametogenesis, such as certain imprints, cannot be restored after nuclear transfer experiments. The fact that animals derived from nuclear transfer survive through birth and beyond suggests that a certain level of abnormal epigenetic reprogramming is tolerated during mammalian development. Although cloned animals may appear normal, the data suggest that these animals have an epigenome that is more similar to the adult donor cells than is the epigenome of IVF-derived progeny.

1.7
Aging

The stability of differentiated cells is an essential feature of higher organisms. Specialized post-mitotic cells, such as neurons, or dividing cells, such as fibroblasts or keratinocytes, have uniform, unchanging phenotypes. The question arises whether, during normal aging, some cells acquire altered phenotypes. Changes in DNA methylation could produce such effects. For example, de-novo methylation of a CpG island could switch a gene off, and if it has an important regulatory or cell-specific function, the effect would be deleterious. Similarly, loss of methylation might turn on a gene that is normally inactive. This is known as the ectopic expression of an inappropriate gene. In both situations it is likely that only a small minority of cells in a given tissue would be affected, which makes it difficult to assess the possible contribution of such cells to the overall senescent phenotype.

So far, the evidence for methylation changes during the aging of organisms is somewhat inconsistent. However, age-related methylation changes have been documented in IGF2 [76] and in N33 and Myc in colon cancer [77]. E-cadherin has been identified as methylated in normal bladder tissues from elderly patients [78], the *tau* gene in human cortex tissue [79], and the estrogen receptor in cardiovascular tissue

[80]. It should be remembered that global DNA methylation must be distinguished from specific methylation or demethylation of CpG islands in promoter regions. Some evidence suggests that global methylation declines during aging [81], but CpG island methylation may increase in some genes but not in others [82]. Indeed, Issa [82] has suggested that a natural increase in the methylation of tumor-suppressor genes during aging may predispose cells to become neoplastic. One important advance is the development of a method that can detect one methylated gene in a thousand cells [83]. Such methods are necessary for analyzing cell populations that are heterogeneous in their pattern of methylation, as will be required to determine in detail the relative importance of methylation defects in tissues that show clear age-related changes. The problem is compounded by the fact that aging can be associated with changes in proteins, membranes, organelles, DNA sequences, and so on.

Some of the best evidence that epigenetic defects can occur during aging comes from the study of X-chromosome reactivation in mice [84, 85]. Genes on the inactive X chromosome have methylated CpG islands, so the observed reactivation may be attributed to demethylation, but this has not been tested directly. The reactivation of the inactive X in during aging of human females does not approach the rate seen in mice and may not occur at all [86]. This raises the important point that short-lived animals are likely to have much looser control of gene expression than long-lived ones. This is already established by the fact that cultured mouse and rat cells can spontaneously transform to neoplastic cells, whereas cultured human diploid cells are extremely resistant to such transformation. It is highly probable that human cells, as well as those from other long-lived species, are much more resistant to random epigenetic changes in gene expression than cells from short-lived species such as rodents. More generally, it can be stated with confidence that cell and tissue maintenance is far more effective in long-lived species and also that many different maintenance mechanisms exist [87].

Cultured primary diploid cells gradually lose total DNA methylation during their limited in-vitro lifespan [88]. It is striking that the level of methylation declines most rapidly in mouse cells, most slowly in human cells, and at an intermediate rate in hamster cells. These rates of change relate directly not only to the cells' lifespans (i.e., shortest in mouse cells and longest in human), but also to the longevities of the donor species. A single treatment of young human cells with the demethylating agent azacytidine has a dramatic effect in shortening their subsequent lifespan [89, 90]. Thus, a single treatment appears to alter the methylation "clock" and suggests that the continuing decline in total DNA methylation eventually leads to senescence. This is completely distinct from the currently favored telomere theory of senescence, which proposes that the absence of telomerase in normal somatic cells leads to a progressive shortening of telomeric DNA and, subsequently, a senescent phenotype. Most immortalized transformed or neoplastic cell lines have telomerase activity, so can maintain their telomeres by the normal enzyme, but a minority use an alternative mechanism [91]. Such cells maintain a constant level of global DNA methylation, so some link or connection must exist between the control of total DNA methylation and the loss or maintenance of telomeres [92]. Nothing is known about the mechanism that switches telomerase on or off. An obvious possibility is that transcription of the gene is under methylation control, but there appear to be no published investigations of this. Nor is it known why cells with limited in-vitro lifespans lose methylation at a constant rate, whereas immortalized cells stably maintain a given level of DNA methylation.

Aging affects only the body, or soma, of animals, whereas the germ line is potentially immortal. This means that germ cells must maintain their telomeres and also their normal pattern of DNA methylation. Nevertheless, it is likely that epigenetic defects can occur in the germ line, so the question arises whether such defects can be detected and, at least in some cases, removed or repaired. When methylation is lost at some important regulatory site, recombination at meiosis may be specifically induced at or close to that site. Subsequently, heteroduplex or hybrid DNA including the site would be hemimethylated, and the normal maintenance methyltransferase would restore full methylation [93]. Experimental evidence that such a process can occur has been obtained in the fungus *Ascobolus*, to which both genetic and methylation analysis can be applied [22]. Thus, one function of recombination in the germ line may be to maintain cells in a juvenile methylated state.

Although epigenetic defects resulting from aberrant DNA methylation may be an important component of the overall aging process, this is only one of many cellular defects that occur in various contexts. Experimental evidence about the changes in DNA methylation is limited, but new methods are beginning to reveal de-novo methylation of CpG islands in some specific genes [94].

1.8
The Future

Although we can say that epigenomics originated a quarter of a century ago with the papers of Riggs [7] and Holliday and Pugh [8], their ideas were largely ignored for a long time. Recently we have seen a burst of interest (see Fig. 1.1) and some solid achievements. Enzymes have been discovered that can methylate DNA in both a de-novo and a maintenance manner. The bisulfite technique of selective deamination of cytosines but not methylcytosines has been applied. The relation of DNA methylation and chromatin remodeling has begun to be elucidated. Methylation has been found to be a major contributor to the oncogenic process. The role of methylation in genetic and epigenetic diseases is now established. Methylation is also a prominent player in aging and age-related disease and has been suggested to be involved in the long-term memory process, as well as in a wide range of other processes, including degenerative ailments.

What will happen in the next 30 years? Now that the human genome has been sequenced, the field of epigenomics may well be one of the boom fields of biotechnology.

Many challenges remain to be addressed. The mechanism whereby CpG islands become methylated in cancer and why certain islands become methylated on a background of hypomethylation remain for the most part unanswered. In addition, we still do not fully understand if methylation either directly or indirectly causes tran-

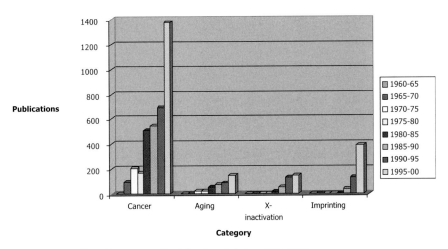

Scientific interest in methylation

Legend:
- 1960-65
- 1965-70
- 1970-75
- 1975-80
- 1980-85
- 1985-90
- 1990-95
- 1995-00

Categories: Cancer, Aging, X-inactivation, Imprinting

Fig. 1.1 Scientific publications in the field of methylation 1960–2000

scriptional silencing in cancers or whether the gene becomes methylated after another inactivation event. Again, is methylation of CpG doublets only important in gene regulation if the doublet is part of a CpG island, or can standalone CpG doublets function as developmental signals too? If these processes can be elucidated and if specific drugs or technologies to reverse the epigenetic cancer phenotype can be found, then this should open up novel specific treatments and cures for cancer. The role of chromatin remodeling is also directly linked to the aberrant methylation patterns observed in cancer. Elucidation of the precise manner in which methylation, methyl-binding proteins, and the histone deacetylation cascade occur will provide new therapeutic targets for the treatment of cancers and other epigenetic-based diseases.

With the use of stem cells for the treatment of disease now just around the corner, the methylation profiles of stem cells will give us valuable insights into the epigenetic regulation that occurs in normal development and in the establishment of tissue specificity, which to date is relatively unknown. The new century has also seen the advent of nuclear cloning. The poor success rate of the technique has been related to incomplete epigenetic reprogramming of the donor nuclei. The study of the epigenetic patterns of critical genes that are incompletely reprogrammed by the oocyte, by efficient methods of sequencing the methylation status of critical regions of genes, could lead to substantial improvements to this technology and to a reduction in fetal abnormalities and associated ailments in the adult animals.

In addition, every person reacts differently to drug treatment. Some people respond well, others have adverse drug reactions–methylation of certain genes as a result of drug treatment may be partly responsible for this phenomenon. Thus in the future, personalized medicines provided as a result of epigenetic profiling of critical genes may be a more effective method of treating patients than the current generic approach.

References

1 R. D. Hotchkiss, *J. Biol. Chem.* **1948**, *168*, 315–332.

2 Y. Gruenbaum, T. Naveh-Many, H. Cedar, A. Razin, *Nature* **1981**, *292*, 860–862.

3 P. Meyer, I. Niedenhof, M. ten Lohuis, *EMBO J.* **1994**, *13*, 2084–2088.

4 A.P. Bird, *Nature* **1986**, *321*, 209–213.

5 P. H. Kay, E. Pereira, S. A. Marlow, G. Turbett, C. A. Mitchell, P. F. Jackson, R. Holliday, J. M. Papadimitriou, *Gene* **1994**, *151*, 89–95.

6 S. J. Clark, J. Harrison, M. Frommer. *Nature Genet.* **1995**, *10*, 20–27.

7 A. D. Riggs, *Cytogenet. Cell Genet.* **1975**, *14*, 9–25.

8 R. Holliday, J. E. Pugh, *Science* **1975**, *187*, 226–232.

9 A. M. Maxam, W. Gilbert, *Meth. Enzymol.* **1980**, *65(1)*, 499–560.

10 G. M. Church, W. Gilbert, *Proc. Natl. Acad. Sci. USA* **1984**, *81*, 1991–1995.

11 H. P. Saluz, J. P. Jost, *Gene* **1986**, *42*, 151–157.

12 G. P. Pfeifer, S. D. Steigerwald, P. R. Mueller, B. Wold, A. D. Riggs, *Science* **1989**, *246*, 810–813.

13 H. Hayatsu, Y. Wataya, K. Kai, S. Iida, *Biochemistry* **1970**, *9*, 2858–2866.

14 G. W. Grigg, *DNA Sequence* **1996**, *6*, 189–198.

15 M. Frommer, C. McDonald, D. S. Millar, C. M. Collis, F. Watt, G. W. Grigg, P. L. Molloy, C. L. Paul, *Proc. Nat. Acad. Sci. USA* **1992**, *89*, 1827–1831.

16 S. J. Clark, J. Harrison, C. L. Paul, M. Frommer, *Nucleic Acids Res.* **1994**, *22*, 2990–2997.

17 G. Grigg, S. Clark, *Genes and Genomes* **1994**, *16*, 431–436.

18 R. P. Paulin, G. W. Grigg, M. W. Davey, A. A. Piper, *Nucleic Acids Res.* **1998**, *26*, 5009–5010.

19 P. M. Warnecke, J. R. Mann, M. Frommer, S. J. Clark, *Genomics* **1998**, *51*, 182–190.

20 P. M. Warnecke, D. Biniszkiewicz, R. Jaenisch, M. Frommer, S. J. Clark, *Dev. Genetics* **1998**, *51*, 182–190.

21 P. M. Warnecke, C. Stirzaker, J. R. Melki, D. S. Millar, C. L. Paul, S. J. Clark, *Nucleic Acids Res.* **1997**, *22*, 111–121.

22 L. Maloisel, J. L. Rossignol, *Genes Dev.* **1998**, *12*, 1381–1389.

23 F. Lyko, B. H. Ramsahoye, R. Jaenisch, *Nature* **2000**, *408*, 538–540.

24 H. Gowher, O. Leismann, A. Jeltsch, *EMBO J.* **2000**, *19*, 6918–6923.

25 R Holliday, *Mutat. Res.* **1991**, *250*, 351–363.

26 N. Federoff, P. Masson, J. A. Banks, *BioEssays* **1989**, *10*, 139–144.

27 L. Rhounim, J. L. Rossignol, G. Faugeron, *EMBO J.* **1992**, *11*, 4451–4457.

28 R. P. Paulin, T. Ho, H. J. Balzer, R. Holliday, *Genetics* **1998**, *149*, 1081–1088.

29 R. Holliday, T. Ho. *Somat. Cell Mol. Genet.* **1991**, *17(6)*, 537–542.

30 J. Nyce, *Somat. Cell Mol. Genet.* **1991**, *17(6)*, 543–550.

31 J. Tasseron-de Jong, J. Aker, H. Der Dulk, P. van de Putte, M. Giphart-Gassler, *Biochem. Biophys. Acta* **1989**, *1008*, 62–70.

32 A. Graessman, G. Sandberg, E. Guhl, N. Graessman, *Mol. Cell Biol.* **1994**, *14*, 2004–2010.

33 P. A. Jones, *Cell* **1985**, *40*, 485–486.

34 D. V. Santi, A. Norment, C. E. Garrett, *Proc. Natl. Acad. Sci. USA* **1984**, *81*, 6993–6997.

35 M. Stam, J. N. K. Mol, J. M. Kooter, *Ann. Bot.* **1997**, *79*, 3–12.

36 F. Linn, I. Heidmann, H. Saedler, P. Meyer, *Mol. Gen. Genet.* **1990**, *222*, 329–326.

37 E. J. Finnegan, R. K. Genger, W. J. Peacock, E. S. Dennis, *Annu. Rev. Plant Mol. Biol.* **1998**, *49*, 223–247.

38 T. Bestor, *Nucleic Acids Res.* **1987**, *15*, 3835–3843.

39 R. HOLLIDAY, *J. Theor. Biol.* **1991**, *151*, 351–358.

40 V. B. MORRIS, R. COWAN, *Cell Tissue Kinet.* **1984**, *17*, 199–208.

41 R. HOLLIDAY, *Biol. Rev. Camb. Philos. Soc.* **1990**, *65*, 431–471.

42 M. MONK, M. BOUBELIK, S. LEHNERT, *Development* **1987**, *99*, 371–382.

43 A. RAZIN, H. CEDAR: *DNA Methylation: Molecular Biology and Biological Significance*, Birkhauser, Basel **1993**, 343–357.

44 E. LI, T. H. BESTOR, R. JAENISCH, *Cell* **1992**, *69*, 915–926.

45 M. OKANO, D. W. BELL, D. A. HABER, W. LI, *Cell* **1999**, *99*, 247–257.

46 M. J. DONOGHUE, B. L. PAATTON, J. R. SANES, J. P. MERLIE, *Development* **1992**, *116*, 1101–1112.

47 P. A. JONES, M. J. WOLKOWICS, W. M. RIDEOUT III, F. A. GONZALEZ, C. M. MARZIASZ, G. A. COETZEE, S. J. TAPSCOTT, *Proc. Natl. Acad. Sci. USA* **1990**, *87*, 6117–6121.

48 W. M. RIDEOUT III, P. EVERSOLE-CIRE, C. H. SPRUCK, C. M. HUSTAD, G. A. COETZEE, F. A. GONZALEZ, P. A. JONES, *Mol. Cell Biol.* **1994**, *14*, 6143–6152.

49 J. YISRAELI, R. ADELSTEIN, D. MELLOUL, U. NUDEL, D. YAFFE, H. CEDAR, *Cell* **1986**, *46*, 409–416.

50 D. FRANK, N. LIGHTENSTEIN, Z. PAROUSH, Y. BERGMAN, M. SHANI, A. RAZIN, H. CEDAR, *DNA Methylation and Gene Regulation*, Royal Society, London **1990**, 63–67.

51 J. P. JOST, H. P SALUZ, *DNA Methylation: Molecular Biology and Biological Significance*, Birkhauser, Basel **1993**, 425–451.

52 J. E. BURNS, D. J. BAGNALL, J. D. METZGER, E. S. DENNIS, W. J. PEACOCK, *Proc. Natl. Acad. Sci. USA* **1993**, *90*, 287–291.

53 C. C. SHELDON, E. J. FINNEGAN, D. T. ROUSE, M. TADEGE, D. J. BAGNALL, C. A. HELLIWELL, W. J. PEACOCK, E. S. DENNIS, *Curr. Opinion Plant Biol.* **2000**, *3*, 418–422.

54 S. KOCHANEK, M. TOTH, A. DEHMEL, D. RENZ, W. DOERFLER, *Proc. Natl. Acad. Sci. USA* **1990**, *87*, 8830–8834.

55 A. BEN-KRAPPA, I. HOLKER, U. SANDARADURA DE SILVA, W. DOERFLER, *Genomics* **1991**, *11*, 1–7.

56 R. HOLLIDAY, *Br. J. Cancer* **1979**, *40*, 513–522.

57 M. A. GAMMA-SOSSA, R. M. MIDGETT, V. A. SLAGEL, S. GITHENS, K. C. KUO, C. W. GEHRKE, M. EHRLICH, *Biochim. Biophys. Acta* **1983**, *740*, 212–219.

58 A. P. FEINBERG, B. VOLGELSTEIN, *Nature* **1983**, *301*, 89–92.

59 S. B. BAYLIN, J. W. HOPPENER, A. BUSTROS, P. H. STEENBERGH, C. J. LIPS, B. D. NELKIN, *Cancer Res.* **1986**, *46*, 2917–2922.

60 J. G. HERMAN, F. LATIF, Y. WENG, M. I. LERMAN, B. ZBAR, S. LIU, D. SAMID, D. S. DUAN, J. R. GNARRA, W. M. LINEHAN, *Proc. Natl. Acad. Sci. USA* **1994**, *91*, 9700–9704.

61 C. STIRZAKER, D. S. MILLAR, C. L. PAUL, P. M. WARNECKE, J. HARRISON, P. C. VINCENT, M. FROMMER, S. J. CLARK. *Cancer Res.* **1997**, *57*, 2229–2237.

62 D. S. MILLAR, K. K. OW, C. L. PAUL, P. J. RUSSELL, P. L. MOLLOY, S. J. CLARK, *Oncogene* **1999**, *18*, 1313–1324.

63 J. R. MELKI, P. C. VINCENT, S. J. CLARK, *Cancer Res.* **1999**, *59*, 3730–3740.

64 R. HOLLIDAY, T. HO, *Proc. Natl. Acad. Sci. USA* **1998**, *95*, 8727–8732.

65 J. WEBER, *Cancer Res.* **1983**, *43*, 3466–3492.

66 W. M. RIDEOUT III, D. EGGAN, R. JAENISCH, *Science* **2001**, *293*, 1093–1098.

67 D. HUMPHREYS, D. EGGAN, H. AKUTSU, K. HOCHEDLINGER, W. M. RIDEOUT III, D. BINISZKIEWICZ, R. YANAGIMACHI, R. JAENISCH, *Science* **2001**, *293*, 95–97.

68 N. OGONUKI, K. INOUE, Y. YAMAMOTO, Y. NOGUCHI, K. TANEMURA, O. SUZUKI, H. NAKAYAMA, K. DOI, Y. OHTOMO, M. SATOH, A. NISHIDA, A. OGURA, *Nat Genet.* **2002**, *30(3)*, 253–254.

69 R. P. LANZA, J. B. CIBELLI, C. BLACKWELL, V. J. CRISTOFALO, M. K. FRANCIS, G. M. BAERLOCHER, J. MAK, M. SCHERTZER, E. A. CHAVEZ, N. SAWYER, P. M. LANSDORP, M. D. WEST, *Science* **2000**, *288*, 665–669.

70 K. J. MCCREATH, J. HOWCROFT, K. H. CAMPBELL, A. COLMAN, A. E. SCHNIEKE, A. J. KIND, *Nature* **2000**, *405*, 1066–1069.

71 S. M. TILGHMAN, *Cell* **1999**, *96*, 185–193.

72 Y. K. KANG, D. B. KOO, J. S. PARK, Y. H. CHOI, H. N. KIM, W. K. CHANG, K. K. LEE, Y. M. HAN, *J. Biol. Chem.* **2001**, *276*, 39980–39984.

73 Y. K. KANG, D. B. KOO, J. S. PARK, Y. H. CHOI, K. K. LEE, Y. M. HAN, *FEBS Lett.* **2001**, *499*, 55–58

74 M. TADA, Y. TAKAHAMA, K. ABE, N. NA-KATSUJI, T. TADA, *Curr Biol.* **2001**, *11*, 1553–1558.

75 Y. K. KANG, D. B. KOO, J. S. PARK, Y. H. CHOI, A. S. CHUNG, K. K. LEE, Y. M. HAN, *Nature Genet.* **2001**, *28*, 173–177.

76 J. P. ISSA, P. M. VERTINO, C. D. BOEHM, I. F. NEWSHAM, S. B. BAYLIN, *Proc. Natl. Acad. Sci. USA* **1996**, *93*, 11757–11762.

77 N. AHUJA, Q. LI, A. L. MOHAN, S. B. BAYLIN, J. P. ISSA, *Cancer Res.* **1998**, *58*, 5489–5494.

78 D. M. BORNMAN, S. MATHEW, J. ALS-RUHE, J. G. HERMAN, E. GABRIELSON. *Amer. J. Pathol.* **2001**, *159*, 831–835.

79 H. TOHGI, K. UTSUGISAWA, Y. NAGANE, M. YOSHIMURA, M. UKITSU, Y. GENDA. *Neurosci. Lett.* **1999**, *275*(2), 89–92.

80 W. S. POST, P. J. GOLDSCHMIDT-CLER-MONT, C. C. WILHIDE, A. W. HELDMAN, M. S. SUSSMAN, P. OUYANG, E. E. MILLIKEN, J. P. ISSA. *Cardiovasc. Res.* **1999**, *43*, 985–991.

81 V. L. WILSON, R. A. SMITH, S. MA, R. G. CUTLER, *J. Biol. Chem.* **1987**, *262*, 9948–9951.

82 J. P. ISSA, *Crit. Rev. Oncol. Hematol.* **1999**, *32*, 31–43.

83 J. G. HERMAN, J. R. GRAFF, S. MYOHA-NEN, B. D. NELKIN, S. B. BAYLIN, *Proc. Natl. Acad. Sci. USA* **1996**, *93*, 9821–9826.

84 B. L. CATTANACH, *Genet. Res.* **1974**, *23*, 291–306.

85 K. A. WAREHAM, M. F. LYON, P. H. GLENISTER, E. D. WILLIAMS, *Nature* **1987**, *327*, 725–727.

86 B. R. MIGEON, M, SCHMIDT, J. AXEL-MAN, C. R. CULLEN, *Proc. Natl. Acad. Sci. USA* **1986**, *83*, 2182–2186.

87 R. HOLLIDAY, *Understanding Ageing*, Cambridge University Press, Cambridge 1995.

88 V. L. WILSON, P. A. JONES, *Science* **1983**, *220*, 1055–1057.

89 R. HOLLIDAY, *Exp. Cell Res.* **1986**, *166*, 543–552.

90 S. FAIRWEATHER, M. FOX, P. MARGISON, *Exp. Cell. Res.* **1987**, *168*, 153–159.

91 R. R. REDDEL, *Ann. N. Y. Acad. Sci.* **1998**, *854*, 8–19.

92 R. HOLLIDAY, *Stem Cell Biology*, Cold Spring Harbor Laboratory Press, Woodbury, New York 2001, 95–109.

93 R. HOLLIDAY, *Science* **1987**, *238*, 163–170.

94 J. P. ISSA, *Ann. N. Y. Acad. Sci.* **2000**, *910*, 140–153.

2
(Epi)genetic Signals: Towards a Human Genome Sequence of All Five Nucleotides

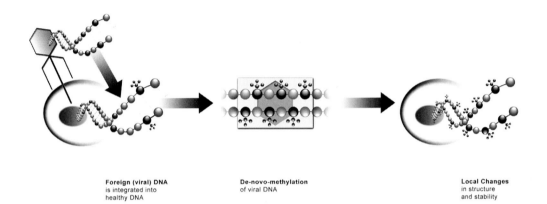

Foreign (viral) DNA
is integrated into
healthy DNA

De-novo-methylation
of viral DNA

Local Changes
in structure
and stability

2
(Epi)genetic Signals:
Towards a Human Genome Sequence of All Five Nucleotides
WALTER DOERFLER*

Summary

Our observations in the late 1970s that chromosomally integrated adenovirus (Ad) DNA, as opposed to Ad DNA in the virion or free in infected cells, becomes methylated de novo started a long series of investigations on the biological function of DNA methylation. Aside from 5-methyldeoxycytidine (5-mC) residues in specific 5'-CG-3' dinucleotide positions being frequently associated with long-term promoter inactivation, DNA methylation patterns in mammalian and other genomes are thought to serve a more general function, in that 5-mC residues can modulate DNA–protein interactions positively or negatively. Methylation patterns can thus assume a function in the buildup of complex chromatin structures. On a limited scale, we determined the distribution of 5-mC residues by the genomic sequencing technique in different segments of the human genome. In some instances, these patterns are interindividually conserved, but can be different from cell type to cell type. In the genetically imprinted Prader-Willi/Angelman region on human chromosome 15q11–13, some segments are totally 5'-CG-3'-methylated on the maternally inherited chromosome and unmethylated on the paternal one. In other subsegments of the same region, this difference is less distinct, in yet others not existing at all. Such alternations in methylation patterns between different segments of an imprinted region may signal part of the imprinting message. We have also provided evidence that the integration of foreign – Ad, bacteriophage λ, plasmid – DNA into established mammalian genomes not only leads to de novo methylation of the integrates, but can also elicit changes in cellular methylation and transcription patterns. Thus, insertions of foreign DNA into mammalian genomes have to be critically viewed for possible consequences to the stability of the target genome. This topic is of particular interest in studies on the mechanism of viral oncogenicity. Viral systems will con-

* Acknowledgments: Research in the author's laboratory over the past 30 years was supported at different times by the Deutsche Forschungsgemeinschaft, Bonn (SFB 74, SFB 274, Do-165-15, Do-165-17), the Bundesministerium für Bildung und Wissenschaft, the Thyssen Stiftung, Bonn, the Alexander von Humboldt Stiftung, Bonn, the Sander Stiftung, München, the Ministerium für Wissenschaft und Forschung des Landes Nordrhein-Westfalen, and the Bayerisches Staatsministerium für Landschaftsschutz und Umweltgestaltung.

tinue to serve as interesting models for many important aspects of biological DNA methylation. In this context, it is worth recalling that the DNA of the iridovirus frog virus 3 (FV3) is completely 5'-CG-3'-methylated. Completely methylated FV3 promoters are active in both fish and mammalian cells. At this time, investigations on the significance of DNA methylation, although still far from answering all the intriguing questions, promise interesting avenues of exciting research on the structure and function of complex genomes.

2.1
A Linguistic Prologue

Having worked on DNA methylation since 1975, one of the best-investigated factors among epigenetic mechanisms, I still wonder whether the term "epigenetic" should be used when addressing an important modulating signal in genetics. Since words and their meanings have an intimate relationship to human history, they can be complicated and confusing when we use them perhaps too lightly. In preparing to write this chapter, I consulted my Greek dictionary of highschool days. Numerous entries under the Greek preposition "επι", among them *on, at, in, close to, during, over, next to*, are applicable in both temporal and spatial senses. Of course, a methyl group is added post-replicationally onto cytidine residues in a DNA sequence. Hence the term "επι" might have its biochemical justification. However, in Greek literature, "επι" has also had the flavor – according to my dictionary – of *dominance over, dependence on, under someone's power*. Do we then want to imply that all of genetics is *under the dominance* of methyl groups distributed in the genome in highly significant patterns? Some geneticists would probably not let us get away with this interpretation. Nevertheless, the authors of the chapters collected in this book are united by their daily scientific experience that the fifth nucleotide positioned at genetically crucial sites does indeed exert its *dominance over* functionally important DNA–protein interactions and thus decisively influences most genetic mechanisms. I would, therefore, like to argue that a signal, which is capable of affecting many genetic mechanisms so fundamentally, better be understood as a *bona fide* genetic signal. The term epigenetic would suggest a mechanism outside the range of regular genetic processes when, in fact, 5-mC can play a crucial role right in the central arena of genetics.

2.2
Towards the Complete Sequence of the Human Genome with Five Nucleotides

A realistic general way of looking at patterns of DNA methylation has to be based on their ability to modulate DNA–protein interactions both positively and negatively. Thus, a 5-methyldeoxycytidine (5-mC) residue in a particular position of the human genome constitutes the primary modulating signal and influences the interactions of this DNA segment with proteins in the construction of chromatin modules. The primary array of proteins being attracted to or repulsed from interacting with specific DNA sequences that are methylated, hypomethylated, or unmethylated in specific patterns then forms the scaffold for the binding of additional proteins and their interactions with subsequent layers of proteins, which thus contribute to the buildup of specific chromatin structures in the human genome. This model may explain one of the factors affecting the emergence of specific chromatin configurations. However, barring the availability of a complete map of all the 5-mC groups at least in major parts of the human genome, it will be impossible to propose more specific models. Moreover, it is likely that the specific distribution of 5-mC residues could assume quite different functions, particularly when comparing highly functional sequence segments, for example, in a promoter sequence as opposed to repetitive DNA elements with completely unknown functions. Nevertheless, the information on the position of a single 5-mC residue or on the methylation pattern in a promoter sequence or even in an entire gene will not provide nearly enough information to understand the biological significance of the overall pattern which a single 5-mC is only a part of. If for no other reason, it will be extremely important to obtain the complete sequence of the human genome, and the complete sequence will have to include the fifth nucleotide 5-mC. Here we have the paramount justification for the genomic sequencing program for the human genome, which will encompass all 5 nucleotides: A, C, G, T – and 5-mC.

An amazingly large number of clinicians and even biologists are unaware of or, for different reasons, incapable of recognizing the existence and functional importance of the fifth nucleotide. Hence, the Human Epigenome Project will also have to shoulder a major public relations task, informing the general public.

2.3
Patterns of DNA Methylation – the Scaffold for Building a Functional Genome

The prime goals of the human (epi-)genomic sequencing program will be (i) to evaluate the feasibility of and (ii) to develop methods for assessing the distribution of 5-mC residues in the entire human genome. The much-acclaimed human genome projects of the past have failed to recognize the importance of the fifth nucleotide and have not attempted to contribute to the complex technology required to determine 5-mC in the nucleotide sequence. The methods presently available to this end are still time-, cost-, and labor-intensive, although very precise. Since each organism, and probably each cell type in a given organism, has its characteristic pattern of 5-mC distribution which is specific for each genome segment, any technique to be developed for establishing these patterns must avoid – at least initially – the conventional cloning of DNA fragments, which was the basis of the human genome projects of the past. For this purpose, genomic sequencing methods were invented [1–3], which work according to the following principles:

1. Upon careful denaturation, the entire genomic DNA is treated with a chemical (hydrazine or bisulfite) that converts C, but not 5-mC, to uracil. Hence, all 5-mC nucleotides remain in their authentic positions in the sequence.

2. The genome segments to be analyzed are next amplified by the polymerase chain reaction (PCR), and the PCR products are cloned into bacterial plasmid vectors. At this stage, all 5-mC residues are transcribed as C residues; the original C residues converted in step 1 to uracils are, however, transcribed as T residues.
3. The nucleotide sequences of at least 10 clones are determined and compared with the previously established nucleotide sequence of the segment being studied.
4. Controls, to verify the conversion of all C residues to T residues and to ascertain the faithfulness of the PCR conditions, include analyses of unmethylated or of *in vitro* 5-mC premethylated nucleotide sequences of the known DNA segments by the same methods.

For the past two decades, this approach has been the only reliable method of determining patterns of DNA methylation, i.e., the distribution of 5-mC residues in parts of the human and other genomes. A meaningful analysis of DNA methylation patterns has therefore to take into account:

- That all 5-mC residues must be found in their authentic sequence positions.
- That these patterns can be different for each genome segment, each cell type, and each organism.
- That one cannot be certain of the stability of these patterns under conditions of flexible functionality of the genome.
- That fundamental changes in methylation patterns occur during embryonic and fetal development of organisms, in tumor cells, and in human diseases. It is likely that the patterns of DNA methylation are subject to alterations depending on the various functional states of a cell.

There are interesting proposals in the literature about the nature, meaning, and conservation of DNA-methylation patterns. Most of these purely hypothetical notions have been based on analyses of DNA segments of limited length and have most frequently been derived from the determination of these patterns with methods of limited resolution, e.g., by using methylation-sensitive restriction enzymes. Hence, for the purpose of this discussion on how to launch a human genomic sequencing program, it will be advisable to start all over again and not to assume that we understand the detailed functional significance of DNA methylation patterns.

2.4
DNA Methylation Patterns in Segments of the Human Genome and in Viral Genomes

By using the hydrazine [1] and later the bisulfite protocol of the genomic sequencing method [2, 3], we have started to determine the distribution of 5-mC residues in various segments of the human genome. These projects have been connected to specific functional questions that were frequently pursued in the context of medically oriented molecular analyses. Our main motivation for these studies has, however, been the quest to elucidate patterns of DNA methylation in the human genome. Of course, patterns of 5-mC residues in large parts of the human genome could not possibly be determined with the resources of a single laboratory. The segments analyzed in our laboratory have been selected nearly at random, although they often stem from gene or promoter sequences. Due to the longstanding interest of our laboratory in molecular virology, the analyses of several viral genomes, both free or in the integrated state, have also been included. We therefore reasoned that it would be useful to start with an exemplary approach linked to functional aspects and, depending on the results of these studies, to document the need for a more comprehensive investigation of methylation patterns in the entire human genome. In this chapter, I summarize some of our results obtained so far (Table 2.1).

Since our data are necessarily limited, only preliminary interpretations can be derived from them. The literature on DNA methylation abounds with generalizations on patterns of DNA methylation and their possible significance. Some of the "rules" proposed there are probably not generally applicable and frequently are based on analyses of DNA methylation by the HpaII/MspI restriction enzyme approach. In most DNA sequences, the 5'-CCGG-3' recognition sequence constitutes less than 10%–15% of all 5'-CG-3' dinucleotides. I will proceed with caution and with an unavoidable bias towards the validity of genomic sequencing data and comment in what follows on some of our observations.

2.4.1
On Viral Genomes and Foreign DNA Integrates (Table 2.1)

- The virion genomes of the human adenoviruses and many other mammalian viruses do not contain 5-mC residues, i.e., the DNA encapsidated into virus particles is devoid of 5-mC [6]. The free Ad2 viral DNA in infected cells is not methylated either [19].
- In contrast, the genome of the iridovirus frog virus 3 (FV3) in the virion, and also in infected fish or mammalian cells soon after viral DNA replication, is probably methylated at all 5'-CG-3' sequences [11].
- Even fully 5'-CG-3'-methylated frog virus 3 promoters in the process of transcription are active. In reconstruction experiments, one of these promoters, that of the *L1140* gene, lost activity when only the eight 5'-CCGG-3' (HpaII) sites were methylated [20]. Each promoter may have its specific methylation pattern which is compatible with the active or inactive state. Apparently, a specific 5'-CCGG-3' methylation profile not naturally found in this FV3 promoter is not consistent with its activity.
- When adenovirus genomes become part of the cellular genome upon their integration, they are methylated *de novo* in specific patterns that are related to the states of activity of the viral genes in transformed and tumor cells [7–9]. These patterns of DNA methylation and viral gene activity are probably, at least in part, determined by the transformed or tumor phenotype of the cells that carry the integrated viral genomes.
- The *de novo* methylation of integrated foreign (adenovirus, bacteriophage λ, plasmid) DNA in mammalian cells is initiated internally in the integrates in several regions [9]. At the nucleotide level, the site of initiation within such a region ap-

Tab. 2.1 Segments of the human genome or of viral genomes in which the distribution of 5-mC residues was determined by genomic sequencing methods or by restriction analyses

DNA segment (cell type)	Length of DNA sequence in nucleotide pairs (number of 5'-CG-3')[a]	# of 5-mC's	Chromosomal location	Reference
Interleukin-2 Receptor Alpha Chain (IL-2Rα) (lympho-, granulocytes)	600 (15) 2.5%	(1)	10p14–15	[4]
Alu-α 1 Globin Gene-associated (PWBC, sperm, testis, heart)	about 200 (9) 4.5%	9	16p13	[5]
Tissue-plasminogen Activator Gene, Introns 8, 9 (PWBC, sperm, brain, testis)	about 300 (12 HpaII, HhaI sites)	12	8p12	[5]
Alu-ACTH Gene-associated (PWBC, sperm, brain, heart, testis)	about 300 (22) 7.3%	0	1q23–24	[5]
Alu-angiogenin Gene-associated (PWBC, sperm, brain, testis, heart)	about 300 (cluster of CGs)	6 HpaII; HhaI sites; 7 CG sites (unmethylated in sperm)	14q11.1–11.2	[5]
Adenovirus Type 2 (virion DNA)	35.937 (2429) 6.76%	0	–	[6]
Adenovirus Type 12 (virion DNA)	34.125 (1500) 4.40%	0	–	[6]
Adenovirus Type 12 integrated in the hamster genome)	Heavily methylated in specific patterns. *De novo* methylation is initiated in several internal regions of the integrated Ad12 DNA.		Many different integration sites	[7–10]
Frog Virus 3 (iridoviridae)	Complete sequence not available	Completely methylated	–	[11]
TNF-α Gene (granulo-, monocytes) (T-cells, NK cells)	800 (25) 800 (25) 3.1%	3 1–2	6p21.3	[12]
TNF-ß gene (granulo-, monocytes) (T and B cells)	200 (13) 6.5%	13	6p21.3	[12]
β-spectrin Gene (PWBC)	450 (47) 10.4%	0	14q23–24.2	[13]
Protein 4.2 Gene (PWBC)	about 940 (10) 1.1%	11 (50–100%)[b]	15q15–21	[13]

Tab. 2.1   (continued)

DNA segment (cell type)	Length of DNA sequence in nucleotide pairs (number of 5'-CG-3')[a]	# of 5-mC's	Chromosomal location	Reference
Band 3 Gene (PWBC)	450 (15) 3.3%	15 (33–100%)[b] two positions (0–70%)	17q12-ter	[13]
FMR1 Gene Promoter and 5' Region (PWBC, healthy individuals)	about 340 (49) 14.4%	0	Xq 27.3	[14]
FMR1 Gene Promoter (PWBC, FraXA patients)	The expanded 5'-(CGG)-3' repeat is heavily but variably methylated	variable	Xq 27.3	[14]
RET Protooncogene Promoter (PWBC)	400 (49) 12.2%	0	10q11.2	[15]
PWS/AS[c] Region on Human Chromosome 15: Maternally Inherited Chromosome			15q11–13	
	SNRPN segment 430 (23) 5.3%	23		[16]
	PWCFOA segment 653 (7) 1.1%	7 (6)[d]		[16]
	AS-SRO 800 (12) 1.5%	12		[17]
Paternally Inherited Chromosome	SNRPN segment 430 (23) 5.3%	0		[16]
	PWCFOA segment 653 (7) 1.1%	1 (2)[e]		[16]
	AS-SRO 800 (12) 1.5%	12		[17]
540 kb of *randomly selected human* DNA (HpaII/MspI analyses only)	Heavily methylated, identical patterns among different individuals of different ethnic origins. Unknown sequences.	Only HpaII sites analyzed, heavily methylated.	–	[18]

[a] The percentages refer to the number of C residues in 5'-CG-3' dinucleotides in a given nucleotide sequence.
[b] Methylation was variable; some clones were unmethylated.
[c] Imprinted region.
[d] Unmethylated clones were found.
[e] Methylated clones were found.

pears to be variable and not specific for a certain nucleotide position [10]. DNA methylation then spreads, starting from the sites of initiation, in both directions to major parts of the genome [21, 22].

- The "rules" of *de novo* methylation are not understood. Among the recognizable factors that affect *de novo* methylation are (i) the sites of integration of foreign

DNA in established mammalian genomes [9]; (ii) possibly the strengths of promoters, in that weak promoters tend to become methylated, and strong promoters tend to be more resistant to *de novo* methylation [23]; (iii) the nucleotide sequence of the integrate may also contribute to the regulation of *de novo* methylation.

- Since the expansion of the 5'-CGG-3' repeats in the 5'-upstream region of the FMR1 gene on human chromosome Xq27.3 leads to their (variable) *de novo* methylation, it is conceivable that these expanded nucleotide sequences are recognized as foreign [24] and hence become methylated *de novo* like any other foreign DNA, e. g., integrated viral DNA [14].

- Since integrated foreign genomes, like that of the adenoviruses, are methylated *de novo* in distinct but not uniform patterns, it is unrealistic to argue that any nucleotide sequence in foreign DNA would become methylated *de novo*. There is an element of selectivity, which is probably related to the functional states of the integrated genes and other unknown factors. Virus-transformed mammalian tumor cells may constitute a particularly interesting, functionally selected class of cells with a specific regulation of *de novo* methylation.

2.4.2
DNA Methylation Patterns in the Human Genome (Table 2.1)

- The density of 5'-CG-3' residues in a nucleotide sequence or the state of activity of a promoter is not a sufficient parameter for predicting the presence or absence of 5-mC residues. The nucleotide sequences in the (active) β-spectrin [13] or the (inactive) RET promoter [15] contain 10.4% and 12.2% 5'-CG-3' dinucleotides, respectively, but are both completely unmethylated in peripheral white blood cells (PWBC). In contrast, the TNF-β gene promoter with 6.5% 5'-CG-3' dinucleotides is completely methylated in PWBC. In the protein 4.2 gene expressed in the erythrocyte membrane, the 1.1% of 5'-CG-3' dinucleotide sequences are 50%–100% methylated [13]. In the TNF-alpha promoter with 3.1% 5'-CG-3' dinucleotides, 3 out of 25 are methylated, in the TNF-β promoter with 6.5% 5'-CG-3' dinucleotides, all 13 CGs are methylated in the same cell types [12, 25].

- Active promoters in the human genome are frequently hypo- or unmethylated. The C residues in a promoter, the methylation status of which co-determines the activity of the promoter [26], can be different in each promoter. Most probably, the decisive sequences in a promoter are those that interact with specific transcription factors that depend on the presence or absence of 5-mC residues. Hence, the methylation status of a promoter sequence by itself does not allow predictions about its activity.

- Inactive promoters in the human genome can be heavily methylated or even completely unmethylated. We surmise, but cannot prove this notion, that promoters that need to be reactivated occasionally and are not permanently shut down remain unmethylated or hypomethylated. The influence of other factors must be important. The promoters of genes that are active only during embryonic and/or fetal development and inactive in adults may be the highly methylated ones.

- Imprinted segments of the human genome, like the Prader-Labhart-Willi/Angelman region on human chromosome 15q11−13, exhibit strictly differential methylation patterns only in parts of the region [16, 17]. The data in Table 2.1 demonstrate that in the SNRPN segment all 23 5'-CG-3' dinucleotides in a sequence of 430 nucleotides are fully methylated on the maternally inherited chromosome but completely unmethylated on the paternally inherited chromosome. However, in the PWCFOA segment, which is located many kilobases centromeric from the SNRPN region, the pattern differences between the two chromosomes are less unequivocal, and in the AS-SRO region, methylation on both chromosomes is practically identical. Hence, it is by no means clear that all parts of an imprinted chromosomal segment differ in the extent of 5'-CG-3' methylation on the two chromosomes. Perhaps these variations in the intensity of methylation in different segments of an imprinted genome segment are part of the functional imprinting message.

- A given pattern of promoter methylation, e.g., that in the TNF-alpha promoter in granulo- and monocytes, can be interindividually highly conserved and be independent of the ethnic origin of the probands [12, 25].
- The methylation status of promoter sequences can be different in different cell types, possibly also at different times. Hence, the determination of methylation patterns in promoter sequences will be a complicated, laborious undertaking. Without taking into account these (unpleasant for the investigator) facts, the determination of patterns will remain of limited value.
- The results of genomic sequencing analyses encompassing all 5'-CG-3' sequences in various segments of the human genome do not suggest simple relationships between the density of 5'-CG-3' dinucleotides in a genome segment and the extent of their methylation. Even imprinted segments exhibit parts with varying levels of DNA methylation, and the expanded 5'-$(CGG)_n$-3' repeat in the 5'-upstream region of the *FMR1* gene on human chromosome Xq27.3 exhibits highly variable DNA methylation in FraXA patients [14].

My intention in presenting these data is to recommend caution – particularly to myself – in drawing too simple conclusions about the nature and biological significance of the complex patterns of DNA methylation in the human genome. The inherent and apparent difficulties with the interpretation of these data has convinced me and may help to convince others that we have a highly important but extremely difficult task ahead: the exact assessment of these patterns in large segments of the human genome, in genes, regulatory regions, and, perhaps most importantly, in repetitive segments including retrotransposons.

2.5
Insertions of Foreign DNA into Established Mammalian Genomes

Little is known about the stability of patterns of DNA methylation in the mammalian genome. Alterations have been documented during embryonic development (for review, [27]) and possibly during fetal development, but it is uncertain how patterns remain fixed with alterations in the functional states of parts of the genome. There is a huge literature on changes of DNA methylation patterns in specific genes in many different tumor diseases. It will be interesting to await the determination of similar changes in other human diseases.

My laboratory has pursued problems of foreign, mainly viral, DNA insertions into established mammalian genomes for many years. Various aspects of this research were summarized recently [28, 29] and need not be repeated here. Several important sequelae of foreign DNA insertion might be of importance for the stability and functional integrity of an established mammalian genome. It does not seem to matter what foreign DNA is inserted or whether it is introduced into the mammalian cells by viral infection or by DNA transfection. We have observed the following consequences of foreign DNA insertion:

- *de novo* methylation of the integrates, often in specific, possibly function-dependent patterns [7, 9, 10];
- alterations in patterns of DNA methylation in the immediate vicinity of the integrate and – probably more importantly – also remote from it [30, 31];
- changes in the transcription patterns of many cellular genes, perhaps, in part at least, as a consequence of foreign DNA insertion [32].

We are currently investigating to what extent foreign DNA insertion can lead to perturbations in chromatin structure at the site of foreign DNA integration and at considerable distances from these sites (L. Mangel and W. Doerfler, unpublished work). This problem has not been studied before but poses intricate and important questions since, in general, multiple copies of foreign DNA are inserted as a package, most frequently at a single chromosomal locus.

2.6
De Novo Methylation of Integrated Foreign DNA

We have investigated the problem of *de novo* methylation of integrated foreign DNA in two different experimental systems: (i) integrated Ad12 genomes and (ii) the mouse B lymphocyte tyrosine kinase (*BLK*) gene, which has been reintegrated by homologous or heterologous recombination into the mouse genome [23]. Our current, still limited understanding of problems related to the *de novo* methylation of integrated DNA in these systems is summarized in this section.

2.6.1
Ad12 Genomes in Hamster Tumor Cells

In our first experiments in this project, integrated Ad12 genomes in hamster tumor and transformed cells could not be cleaved by methylation-sensitive restriction enzymes like *Hpa*II or *Hha*I, whereas the methylation-insensitive *Hpa*II isoschizomer *Msp*I cleaved the inserted Ad12 DNA to the same pattern as Ad12 virion DNA [7], which we had shown to be nonmethylated. After this discovery on the *de novo* methylation of integrated foreign DNA in mammalian genomes, we started a series of experiments on the biological function of DNA methylation in mammalian cells. A further discovery established, for the first time, an inverse correlation between the levels of promoter methylation and promoter activity [8, 33]. Soon afterwards, we documented that the sequence-specific methylation of viral promoters leads to the inactivation of these promoters [20, 34–39].

In the years that followed, we investigated DNA methylation patterns in the human genome, because we reasoned that DNA methylation must have a biological function that reaches beyond the long-term shutdown of promoters. We thought that this more general role could be the establishment of chromatin domains in which patterns of DNA methylation might present the initial scaffold on which to build more complex structures by specific DNA-protein interactions. Several human genome segments were therefore investigated for specific DNA methylation patterns. Patterns of methylation were highly specific for different parts of the human genome and were different from cell type to cell type. The patterns of DNA methylation in the human genome were frequently highly conserved between individuals (see Sect. 2.4 and Table 2.1).

We set out to elucidate the establishment and mechanisms of *de novo* methylation by using integrated Ad12 genomes as models. Currently, our results can be summarized as follows: Upon integration of Ad12 DNA into the genome of Ad12-induced tumor cells, *de novo* methylation appears to be initiated at specific sites in the genome. The results gleaned from analyses with the bisulfite genomic-sequencing method indicate that the initiation is not at a specific nucleotide or a narrowly restricted set of nucleotides, but rather in a region of the integrated genome. Moreover, there is considerable variability from tumor to tumor with respect to exactly where within this specified region *de novo* methylation begins [10]. From the site(s) of initiation of *de novo* methylation, this modification spreads gradually and progressively, but not uniformly, across major parts of the Ad12 genome. Moreover, certain parts of the Ad12 genome, particularly those that are actively transcribed in the tumor cell, remain unmethylated or become hypomethylated. Further work is required to understand these mechanisms in depth. We surmise that a specific chromatin-like structure across integrated genomes of foreign derivation, like the Ad12 genome, might have a decisive influence on the initiation and spreading of *de novo* methylation.

2.6.2
De Novo Methylation of Foreign DNA Integrated into the Mouse Genome by Homologous or Heterologous Recombination [23]

In one set of experiments, we reinserted the mouse *BLK* gene into the mouse genome by homologous recombination into its authentic site on one of the chromosome 14 alleles in embryonic stem cells. The previously unmethylated *BLK* gene, which had been cloned and propagated in a methylation-deficient bacterial host, became remethylated in exactly the same pattern as on the unmanipulated mouse alleles. When the *BLK* gene landed, however, at heterologous sites elsewhere in the mouse genome, different patterns of *de novo* methylation were observed. We propose that different sites in the mouse genome might have "memory signals" for the establishment of their specific methylation patterns. These signals could be related to specific chromatin properties at a given genomic site. Furthermore, foreign genes that had been attached to the *BLK* gene, like the luciferase gene under the control of the weak (Ad2E2AL) or the strong (early SV40) promoter plus its enhancer, seemed to be methylated differently depending on promoter strength. Removal of the SV40 enhancer led to hypermethylation also of this construct. Tethering the gene to a weak promoter frequently resulted in hypermethylation. In contrast, linkage of the gene to a strong promoter resulted in hypomethylation or absence of *de novo* methylation. This dependence on promoter strength did not hold for all sites in the genome, but was most clearly observed when the construct had been reinserted by homologous recombination.

2.7
Genome-wide Perturbations in the Mammalian Genome upon Foreign DNA Insertion

We have started to investigate the structural and functional consequences of the insertion of foreign DNA into established mammalian genomes. *De novo* methylation of the integrated DNA and changes in DNA methylation patterns in the recipient genomes at the site of insertion and remote from it have been of particular interest. By using different methods, including the bisulfite genomic-sequencing technique [2, 3], we have documented extensive changes in the patterns of DNA methylation at several cellular sites remote from the loci of insertion of the DNA of Ad12 and lesser changes in cells transgenic for the DNA of bacteriophage λ [31, 40]. Since λ DNA is not transcribed in transgenic mammalian cells, changes in methylation patterns subsequent to foreign DNA insertion are not dependent on foreign gene transcription. It was shown earlier that cellular DNA sequences immediately abutting the foreign DNA integrates also exhibit changes in DNA methylation [30]. No one knows by what mechanisms the insertion of foreign DNA affects the organization and function of the recipient genome. Does the site of foreign gene integration determine where the remote effects occur and is there a critical size for the integrate? The acquisition of many kilobases or even a megabase of inserted DNA may alter chromatin topology and thus influence the function of specific parts of the genome also on neighboring chromosomes that are in contact with the site of foreign DNA integration in the interphase nucleus.

A collection of cellular DNA segments and genes was analyzed and searched for changes in DNA methylation and transcription. The technique of methylation-sensitive representational difference analysis (MS-RDA) [41] is based on a subtractive hybridization protocol after selecting against DNA segments that are heavily methylated and hence rarely cleaved by the methylation-sensitive endonuclease *Hpa*II. Use of the MS-RDA protocol led to the isolation of several cellular DNA segments that were indeed more heavily methylated in λ DNA-transgenic hamster cell lines [32]. By applying the suppressive-subtractive hybridization technique to cDNA preparations from nontransgenic and Ad12-transformed or λ DNA-transgenic hamster cells, several cellular genes with altered transcription patterns were cloned from Ad12-transformed or λ DNA-transgenic hamster cells. Many of the DNA segments with altered methylation, which were isolated by a newly developed amplicon subtraction (MS-AS) protocol [42], and cDNA fragments derived from genes with altered transcription patterns were identified by their nucleotide sequences. These segments stemmed from different parts of the genome. In control experiments, no differences in gene expression or DNA methylation patterns were detectable by these methods among individual nontransgenic BHK21 cell clones.

In one mouse line transgenic for the DNA of bacteriophage λ, hypermethylation was observed in the imprinted *Igf2r* gene in DNA from heart muscle. Two mouse lines transgenic for an adenovirus promoter-indicator gene construct showed hypomethylation in the interleukin 10 (IL10) and Igf2r loci. We conclude that the insertion of foreign DNA into an established mammalian genome can lead to alterations in cellular DNA methylation and transcription patterns. It is conceivable that what genes and DNA segments are affected by these alterations depends on the site(s) of foreign DNA insertion [32].

2.8
Outlook and Recommendations

In today's research in molecular genetics, two types of approaches can be distinguished, both of equally fundamental importance:

- elucidation of the structure of genomes and genes and of their immediate functions;
- the search for new genetic signals, mechanisms, and principles.

Investigations of DNA methylation and its biological significance pursue elements of both approaches. In this chapter, I have delivered a personal account of what I consider important in research on DNA methylation and what ought to be done with respect to an (epi)-genomic sequencing program of the human genome. The presentation of the nucleotide sequence of the human genome [43, 44] has been a major, although incomplete, achievement in biomedical research. The human genome sequence presently in hand completely lacks the fifth nucleotide, 5-mC. Stating this omission is not intended to slight the enormous contributions our colleagues have made in providing this vital information to the biomedical community. A technically more complex task requiring sophisticated *finesse* lies now ahead of us, namely the determination of all 5-mC residues in the human genome, in different cell types, at different stages of functionality and development. I have tried to summarize some of my research experience over the past 27 years on the subject of DNA methylation, which have convinced me that the Human Epigenome Project is one of the most important projects in biomedical research for the future if we want to understand at least elements of the function of the entire genome.

References

1 CHURCH, G. M.; GILBERT, W., Genomic sequencing, *Proc. Natl. Acad. Sci. USA* 81, 1991–1995, **1984**.

2 FROMMER, M.; McDONALD, L. E.; MILLAR, D. S.; COLLINS, C. M.; WATT, F.; GRIGG, G. W.; MOLLOY, P. L.; PAUL, C. L., A genomic sequencing protocol that yields a positive display of 5-methylcytosine residues in individual DNA strands, *Proc. Natl. Acad. Sci. USA* 89, 1827–1831, **1992**.

3 CLARK, S. J.; HARRISON, J.; PAUL, C. L.; FROMMER, M., High sensitivity mapping of methylated cytosines, *Nucleic Acids Res.* 22, 2990–2997, **1994**.

4 BEHN-KRAPPA, A.; DOERFLER, W., The state of DNA methylation in the promoter and exon 1 regions of the human gene for the interleukin-2 receptor alpha chain (IL-2R alpha) in various cell types, *Human Molecular Genetics* 2, 993–999, **1993**.

5 KOCHANEK, S.; RENZ, D.; DOERFLER, W., DNA methylation in the Alu sequences of diploid and haploid primary human cells, *EMBO J.* 12, 1141–1151, **1993**.

6 GÜNTHERT, U.; SCHWEIGER, M.; STUPP, M.; DOERFLER, W., DNA methylation in adenovirus, adenovirus-transformed cells, and host cells, *Proc. Natl. Acad. Sci. USA* 73, 3923–3927, **1976**.

7 SUTTER, D.; WESTPHAL, M.; DOERFLER, W., Patterns of integration of viral DNA sequences in the genomes of adenovirus type 12-transformed hamster cells, *Cell* 14, 569–585, **1978**.

8 SUTTER, D.; DOERFLER, W., Methylation of integrated adenovirus type 12 DNA sequences in transformed cells is inversely correlated with viral gene expression, *Proc. Natl. Acad. Sci. USA* 77, 253–256, **1980**.

9 OREND, G.; KNOBLAUCH, M.; KÄMMER, C.; TJIA, S. T.; SCHMITZ, B.; LINKWITZ, A.; MEYER ZU ALTENSCHILDESCHE, G.; MAAS, J.; DOERFLER, W., The initiation of *de novo* methylation of foreign DNA integrated into a mammalian genome is not exclusively targeted by nucleotide sequence, *J. Virol.* 69, 1226–1242, **1995**.

10 HOHLWEG, U.; HÖSEL, M.; DORN, A.; WEBB, D.; SCHRAMME, A.; CORZILIUS, L.; NIEMANN, A.; HILGER-EVERSHEIM, K.; REMUS, R.; SCHMITZ, B.; BUETTNER, R.; DOERFLER, W., Intraperitoneal dissemination of Ad12-induced primitive neuroectodermal hamster tumors: *De novo* methylation and transcription patterns of integrated viral and cellular genes, Under revision, **2002**.

11 WILLIS, D. B.; GRANOFF, A., Frog virus 3 DNA is heavily methylated at CpG sequences, *Virology* 107, 250–257, **1980**.

12 KOCHANEK, S.; TOTH, M.; DEHMEL, A.; RENZ, D.; DOERFLER, W., Interindividual concordance of methylation profiles in human genes for tumor necrosis factors α and β, *Proc. Natl. Acad. Sci. USA* 87, 8830–8834, **1990**.

13 REMUS, R.; ZESCHNIGK, M.; ZUTHER, I.; KANZAKI, A.; WADA, H.; YAWATA, A.; MUIZNIEKS, I.; SCHMITZ, B.; SCHELL, G.; YAWATA, Y.; DOERFLER, W., The state of DNA methylation in promoter regions of the human red cell membrane protein (band 3, protein 4.2 and β-spectrin) genes, *Gene Func. Dis.* 2, 171–184, **2001**.

14 GENÇ, B.; MÜLLER-HARTMANN, H.; ZESCHNIGK, M.; DEISSLER, H.; SCHMITZ, B.; MAJEWSKI, F.; VON GONTARD, A.; DOERFLER, W., Methylation mosaicism of 5'-(CGG)$_n$-3' repeats in fragile X, premutation and healthy individuals, *Nucleic Acids Res.* 28, 2141–2152, **2000**.

15 MUNNES, M.; PATRONE, G.; SCHMITZ, B.; ROMEO, G.; DOERFLER, W., A 5'-CG-3'-rich region in the promoter of the transcriptionally frequently silenced RET protooncogene lacks methylated cytidine residues, *Oncogene* 17, 2573–2584, **1998**.

16 ZESCHNIGK, M.; SCHMITZ, B.; DITTRICH, B.; BUITING, K.; HORSTHEMKE, B.; DOERFLER, W., Imprinted segments in the human genome: different DNA methylation patterns in the Prader-Willi/Angelman syndrome region as determined by the genomic sequencing method, *Human Molecular Genetics* 6, 387–395, **1997**.

17 SCHUMACHER, A.; BUITING, K.; ZESCH-NIGK, M.; DOERFLER, W.; HORSTHEM-KE, B., Methylation analysis of the PWS/AS region does not support an enhancer competition model of genomic imprinting on human chromosome 15, *Nature Genet.* 19, 324–325, **1998**.

18 BEHN-KRAPPA, A.; HÖLKER, I.; SANDA-RADURA DE SILVA, U.; DOERFLER, W., Patterns of DNA methylation are indistinguishable in different individuals over a wide range of human DNA sequences, *Genomics* 11, 1–7, **1991**.

19 KÄMMER, C.; DOERFLER, W., Genomic sequencing reveals absence of DNA methylation in the major late promoter of adenovirus type 2 DNA in the virion and in productively infected cells, *FEBS Letters* 362, 301–305, **1995**.

20 MUNNES, M.; SCHETTER, C.; HÖLKER, I.; DOERFLER, W., A Fully 5'-CG-3' but not a 5'-CCGG-3' methylated late frog virus promoter retains activity, *J. Virol.* 69, 2240–2247, **1995**.

21 TOTH, M.; LICHTENBERG, U.; DOERFLER, W., Genomic sequencing reveals a 5-methylcytosine-free domain in active promoters and the spreading of preimposed methylation patterns, *Proc. Natl. Acad. Sci. USA* 86, 3728–3732, **1989**.

22 TOTH, M.; MÜLLER, U.; DOERFLER, W., Establishment of *de novo* DNA methylation patterns: transcription factor binding and deoxycytidine methylation at CpG and non-CpG sequences in an integrated adenovirus promoter, *J. Mol. Biol.* 214, 673–683, **1990**.

23 HERTZ, J.; SCHELL, G.; DOERFLER, W., Factors affecting *de novo* methylation of foreign DNA in mouse embryonic stem cells, *J. Biol. Chem.* 274, 24232–24240, **1999**.

24 DOERFLER, W., Patterns of DNA methylation–evolutionary vestiges of foreign DNA inactivation as a host defense mechanism–a proposal, *Biol. Chem. Hoppe-Seyler* 372, 557–564, **1991**.

25 KOCHANEK, S.; RADBRUCH, A.; TESCH, H.; RENZ, D.; DOERFLER, W., DNA methylation profiles in the human genes for tumor necrosis factors α and β in subpopulations of leukocytes and in leukemias, *Proc. Natl. Acad. Sci. USA* 88, 5759–5763, **1991**.

26 DOERFLER, W., DNA methylation and gene activity, *Annu. Rev. Biochem.* 52, 93–124, **1983**.

27 RAZIN, A.; SHEMER, R., DNA methylation in early development, *Human Molecular Genetics* 4, 1751–1755, **1995**.

28 DOERFLER, W., A new concept in (adenoviral) oncogenesis: integration of foreign DNA and its consequences, *BBA Rev. Cancer Res.* 1288, 243–244, **1996**.

29 DOERFLER, W.: *Foreign DNA in Mammalian Systems*, Wiley-VCH, Weinheim, New York, 2000.

30 LICHTENBERG, U.; ZOCK, C.; DOERFLER, W., Integration of foreign DNA into mammalian genome can be associated with hypomethylation at site of insertion, *Virus Res.* 111, 335–342, **1988**.

31 HELLER, H.; KÄMMER, C.; WILGENBUS, P.; DOERFLER, W., Chromosomal insertion of foreign (adenovirus type 12, plasmid, or bacteriophage λ) DNA is associated with enhanced methylation of cellular DNA segments, *Proc. Natl. Acad. Sci USA* 92, 5515–5519, **1995**.

32 MÜLLER, K.; HELLER, H.; DOERFLER, W., Foreign DNA integration: genome-wide perturbations of methylation and transcription in the recipient genomes, *J. Biol. Chem.* 276, 14271–14278, **2001**.

33 VARDIMON, L.; NEUMANN, R.; KUHLMANN, I.; SUTTER, D.; DOERFLER, W., DNA methylation and viral gene expression in adenovirus-transformed and -infected cells, *Nucleic Acids Res.* 8, 2461–2473, **1980**.

34 VARDIMON, L.; KRESSMANN, A.; CEDAR, C.; MAECHLER, M.; DOERFLER, W., Expression of a cloned adenovirus gene is inhibited by *in vitro* methylation, *Proc. Natl. Acad. Sci. USA* 79, 1073–1077, **1982**.

35 KRUCZEK, I.; DOERFLER, W., Expression of the chloramphenicol acetyltransferase gene in mammalian cells under the control of adenovirus type 12 promoters: effect of promoter methylation on gene expression, *Proc. Natl. Acad. Sci. USA* 80, 7586–7590, **1983**.

36 LANGNER, K.-D.; VARDIMON, L.; RENZ, D.; DOERFLER, W., DNA methylation of three 5' C-C-G-G 3' sites in the promoter and 5' region inactivates the

E2a gene of adenovirus type 2, *Proc. Natl. Acad. Sci. USA* 81, 2950–2954, **1984**.

37 LANGNER, K.-D.; WEYER, U.; DOERFLER, W., Trans effect of the E1 region of adenoviruses on the expression of a prokaryotic gene in mammalian cells: resistance to 5′-CCGG-3′ methylation, *Proc. Natl. Acad. Sci. USA* 83, 1598–1602, **1986**.

38 WEISSHAAR, B.; LANGNER, K.-D.; JÜTTERMANN, R.; MÜLLER, U.; ZOCK, C.; KLIMKAIT, T.; DOERFLER, W., Reactivation of the methylation-inactivated late E2A promoter of adenovirus type 2 by EIA (13S) functions, *J. Mol. Biol.* 202, 255–270, **1988**.

39 MUIZNIEKS, I.; DOERFLER, W., The impact of 5′-CG-3′ methylation on the activity of different eukaryotic promoters: a comparative study, *FEBS Letters* 344, 251–254, **1994**.

40 REMUS, R.; KÄMMER, C.; HELLER, H.; SCHMITZ, B.; SCHELL, G.; DOERFLER, W., Insertion of foreign DNA into an established mammalian genome can alter the methylation of cellular DNA sequences, *J. Virol.* 73, 1010–1022, **1999**.

41 USHIJIMA, T.; MORIMURA, K.; HOSOYA, Y.; OKONOGI, H.; TATEMATSU, M.; SUGIMURA, T.; NAGAO, M., Establishment of methylation-sensitive-representational difference analysis and isolation of hypo- and hypermethylated genomic fragments in mouse liver, *Proc. Natl. Acad. Sci. USA* 94, 2284–2289, **1997**.

42 MÜLLER, K.; DOERFLER, W., Methylation-sensitive amplicon subtraction: a novel method to isolate differentially methylated DNA sequences in complex genomes, *Gene Funct. Dis.* 1, 154–160, **2000**.

43 International Human Genome Sequencing Consortium, Initial sequencing and analysis of the human genome, *Nature* 409, 860–921, **2001**.

44 VENTER, J. G. et al., The sequence of the human genome, *Science* 291, 1304–1351, **2001**.

3
Epi Meets Genomics: Technologies for Finding and Reading the 5th Base

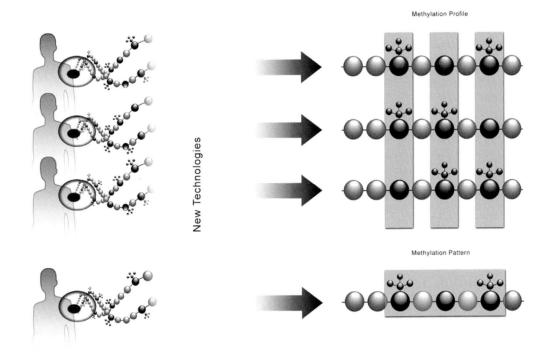

3

Epi Meets Genomics: Technologies for Finding and Reading the 5th Base

Tim Hui-Ming Huang, Christoph Plass, Gangning Liang, and Peter W. Laird

Summary

This chapter presents several high-throughput scanning techniques for the exploration and exploitation of the epigenome. It is now known that the methylation state of multiple 5th bases is susceptible to changes in disease conditions, such as cancer. The described techniques, including restriction landmark genomic scanning (RLGS), methylation-sensitive arbitrarily-primed polymerase chain reaction (MS AP-PCR), and differential methylation hybridization (DMH), generate unique methylation profiles of multiple DNA molecules. Another technique, MethyLight, analyzes methylation patterns of linked CpGs within individual DNA molecules in the genome. The techniques complement one another in the discovery and utilization of this rich resource of epigenome information. RLGS and MS AP-PCR analyses can result in the identification of multiple differentially methylated molecules in disease cells. This large number of candidate sites can further be evaluated by DMH. MethyLight, which is amenable to automated setup and analysis, utilizes the validated 5th-base markers for routine clinical diagnosis.

3.1
The Development of 5th-Base Technologies

The advance of scientific knowledge depends considerably on the development of key technologies. The analysis of methylcytosine, the 5th base in the genome, is no exception. For the past three decades, a number of molecular techniques have been used for analyzing DNA methylation in normal cells and different disease conditions. The application of these methodologies has dramatically increased our understanding of the epigenetic process in the cell as well as aided in the detection and diagnosis of diseases. Reverse-phase high-performance liquid chromatography is an earlier technique used to determine the global content of DNA methylation in different cell or tissue types [1]. However, the discovery of methylation-sensitive restriction enzymes, or endonucleases, and the development of a method based on bisulfite modification of DNA have dramatically advanced progress in this field. These two approaches have established the basis for scientists to develop other analytical tools to study DNA methylation.

3.1.1
Unusual DNA-cutting Enzymes

Restriction enzymes cut DNA into fragments in or near their specific recognition or restriction sites. Interestingly, many enzymes can recognize the same target sequences but cut DNA differently, due to their sensitivity to methylcytosine. These are known as isoschizomers. For example, *Msp*I and *Hpa*II are isoschizomers that recognize the CCGG sequence. When the central CG residue within CCGG is methylated at cytosine, *Hpa*II cannot cleave it, whereas *Msp*I restricts the site regardless the state of methylation of the central cytosine (but is inhibited when the outside C is methylated). The digested DNA fragments can be separated by gel electrophoresis and transferred to a piece of nylon filter that binds DNA, a process called Southern blotting. The nylon filter is then probed with the gene of interest that recognizes specific DNA bands digested by these isoschizomers. If the band patterns differ between the *Hpa*II- and *Msp*I-treated DNA samples, then the DNA sequence of interest likely contains one or more methylated sites. *Hpa*II and *Msp*I were the first pair of enzymes discovered for this type of analysis. Today, many methylation-sensitive restriction enzymes have been identified, some of which do not have isoschizomer partners, and provide valuable resources for research. Southern blotting using DNA cut with different methylation-sensitive enzymes has been a popular method for scientists to study DNA methylation in specific gene sequences (see an example in Fig. 3.1A and B).

3.1.2
A Unique Chemical Reaction that Modifies Methylated DNA

Another commonly used approach is bisulfite modification of DNA, a process that converts cytosine to uracil but leaves the 5th base, methylcytosine, unchanged [2]. When denatured DNA is exposed to sodium bisulfite, a chemical reaction called sul-

---

Fig. 3.1 DNA methylation analysis by Southern hybridization and bisulfite sequencing. **A.** Map location of a *Not*I site located at the 5′ end of the human estrogen receptor α gene. Vertical bars indicate the location of flanking *Eco*RI sites. Pushpins indicate the locations of CpG sites. The pOR3 probe was used as a probe in Southern hybridization shown in **B.** Genomic DNA was digested with *Eco*RI (control) or doubly digested with *Eco*RI and methylation-sensitive *Not*I (MCF-7 and MDA-MB-231 breast cancer cells). The digests were separated by gel electrophoresis and were transferred to a nylon membrane and hybridized with pOR3. This probe identified two DNA fragments (1.9 and 1.2 kb) in MCF-7 cells, indicating that the *Not*I site is not methylated in this cell line. However, this site was methylated in MDA-MB-231 cells, showing that the site was protected from the methylation-sensitive restriction. **C.** Bisulfite DNA was subjected to PCR amplification with primers flanking the *Not*I site and then DNA sequencing analysis. All 4 CG sites (underlined) were unmethylated in MCF-7. In MDA-MB-231, 80% (clone 1) of these sites were completely methylated, and 20% (clones 2 and 3) of the sites were partially methylated.

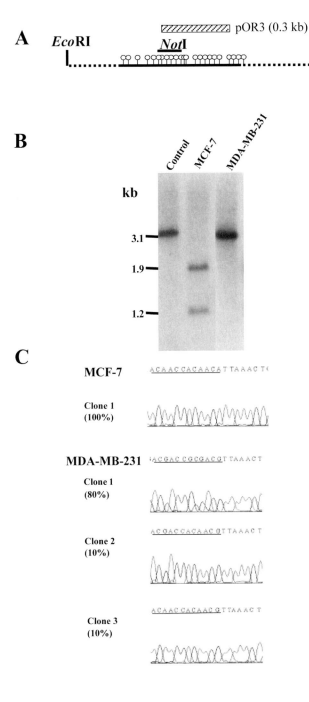

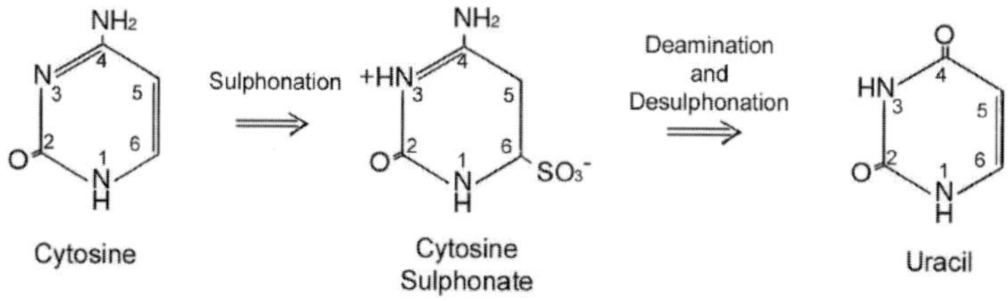

Fig. 3.2 Schematic diagram of bisulfite modification of cytosine

fonation occurs, in which a bisulfite ion (SO_3^-) is added to position 6 of cytosine (Fig. 3.2). In the next step of the reaction, the amino group at position 4 of cytosine sulfonate becomes unstable and deaminates to uracil sulfonate. Subsequently, the SO_3^- adduct is removed from the uracil residue at a high pH. This chemically modified DNA is then amplified by polymerase chain reaction, or PCR, which repeatedly synthesizes a region of DNA located between two primers. Primers are short oligonucleotides that hybridize to opposite strands of a target sequence and prime synthesis of the complementary sequence by the enzyme DNA polymerase. By copying the modified nucleotide sequence into new strands, more than one million prints of the same DNA are made. All uracil and thymine residues are amplified as thymine, and the 5th base, methylcytosine, is amplified as cytosine. The amplified DNA fragments can be processed through nucleotide sequencing, which displays the exact location of 5-methylcytosine in the gene of interest (see an example in Fig. 3.1C).

3.1.3
The Advance of 5th Base Technologies in Epigenomic Research

Using the Southern-based and the bisulfite sequencing methods, work over the past 10 years has led to remarkable discoveries in this field and has provided important insights into the relation of DNA methylation and gene expression. More recently, attention has been drawn to the importance of the role of abnormal DNA methylation in cancer. Alterations in the methylation patterns of the genome are among the most common changes in many types of tumors [3]. DNA hypermethylation, or de-novo methylation, usually occurs by adding a methyl group to the 5th carbon position of a cytosine residue in a base doublet, called a CpG dinucleotide (with the "p" representing the phosphate bond). CpG dinucleotides frequently occur together in regions approximately 1 kilobase long, called CpG islands [4]. These islands are frequently located at the 5′ ends of genes and may participate in the regulation of RNA synthesis initiation, an early transcription event that occurs in the promoter sequence located at the beginning of a gene [5, 6]. DNA hypomethylation, or loss of methylation, is also observed in many cancers and takes place in DNA regions containing repeat sequences [7]. Abnormal DNA methylation has a profound effect on the development of cancer. These important findings lay the groundwork for further development of technologies to analyze complex patterns of DNA methylation present in cancer genomes.

The concept of screening DNA methylation throughout the entire genome becomes increasingly relevant as the Human Genome Project is nearing completion. The challenge is to reap the benefits of the flood of sequence information for further understanding of the epigenetic process in diverse cell types. However, traditional methods, such as Southern blotting and bisulfite sequencing, cannot meet this challenge for genome-wide assays. Scientists must move beyond the notion of "one-gene, one experiment" in favor of highly parallel, automation-based, high-through-put assays.

Inspired by this new concept, our purpose in writing this chapter is to summarize the techniques recently available for global analysis of DNA methylation in complex genomes. These techniques include restriction landmark genomic scanning (RLGS), methylation-sensitive arbitrarily primed PCR (MS AP-PCR), differential methylation hybridization (DMH), and MethyLight. The basic principles and outlines of these techniques and their applications in cancer research are described in the following sections. The chapter is not intended to be inclusive of all techniques [8–12] that are important to advances in this field. For readers who find the material in this chapter too brief, far more detailed accounts of the techniques, along with complete references for other techniques, can be found in the references listed at the end of this chapter.

3.2
Restriction Landmark Genomic Scanning (RLGS): Finding the 5th-base Signposts in the Genomic Atlas

3.2.1
Principle

RLGS is a method that enables searching for changes in DNA methylation in a genome. It uses a combination of DNA-cutting enzymes, one of which is methylation-sensitive. The genomic DNA is digested into distinctive restriction fragments followed by separation of these fragments by two-dimensional gel electrophoresis. RLGS profiles present a display of thousands of restriction fragments tagged with a radioactive tracer (see Fig. 3.3A for an example). Profiles are highly reproducible and thus enable direct comparison between DNAs from two individuals or between normal tissue DNA and tumor DNA profiles (see Fig. 3.3B for an example).

The strength of RLGS is that it can be applied to genomic DNA of any species without prior knowledge of sequence information. The methylation-sensitive restriction enzyme that is used in these experiments is usually a rare cutting enzyme (e. g., at fewer than 5000 sites in the human genome), which preferentially recognizes CpG islands located in the promoter regions of genes. The use of methylation-sensitive restriction enzymes for the RLGS analysis makes it possible to scan genomes for differences in methylation patterns. The restriction landmark enzyme NotI is methylation-sensitive. If a genomic NotI site is methylated, the enzyme does not cut and the site is not labeled with the radioactive tracer, resulting in loss of this fragment

A.

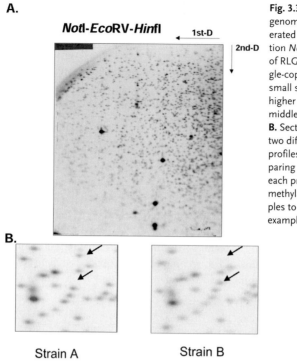

Notl-EcoRV-Hinfl   1st-D

2nd-D

B.

Strain A   Strain B

Fig. 3.3 RLGS profile of mouse genomic DNA. **A.** RLGS profile generated using the enzyme combination Notl-EcoRV-Hinfl. The majority of RLGS fragments represent a single-copy number of genes. Only a small subset of fragments shows higher signal intensity and represents middle-repetitive rDNA sequences. **B.** Section of an RLGS profile from two different mouse strains. RLGS profiles are highly reproducible. Comparing patterns of RLGS fragments in each profile enables differences in methylation status between two samples to be identified (see arrows for examples).

from the RLGS profile. However, if the *Notl* site is unmethylated, the site is cut, the restriction ends are labeled, and the fragment is present in the profile. Since *Notl* sites are mainly located in CpG islands, RLGS is a method that allows determination of the methylation status of thousands of CpG islands at one time.

3.2.2
How Does RLGS Work?

The following steps outline the RLGS procedure (Fig. 3.4):

- **Step 1** – High quality genomic DNA is carefully extracted from fresh-frozen tissue samples. The goal for DNA isolation is to minimize the number of randomly broken DNA fragments whose ends would interfere with the radiolabeling of the restriction landmark sites (see Step 4).
- **Step 2** – The double stranded DNA is "blocked" with nucleotide analogues (blue circles at the ends of DNA fragments in Fig. 3.4), which means that the ends of randomly broken DNA fragments are filled and single-strand overhangs are eliminated. If the number of randomly broken DNA fragments is too large, this step cannot be efficiently completed and results in random labeling of broken DNA ends in Step 4.
- **Step 3** – DNA is digested with a methylation-sensitive rare cutting enzyme such as *Notl*, *AscI*, *BssHII*, or *EagI*. All these enzymes create a 4-base overhang. These

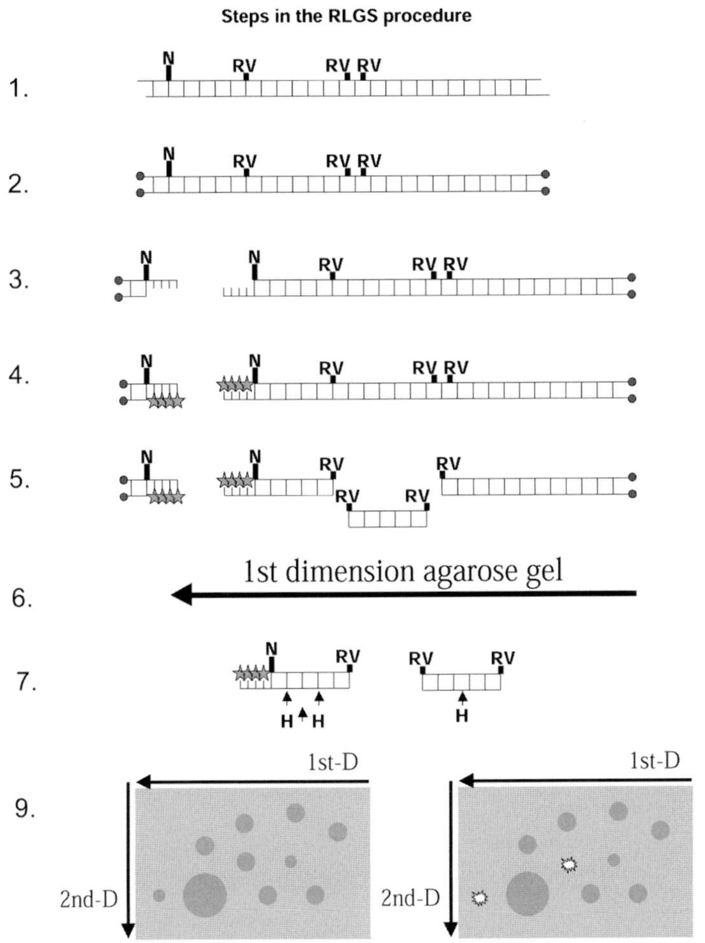

Steps in the RLGS procedure

Fig. 3.4 Schematic outline of the steps involved in RLGS (see text for explanation); restriction enzymes are N-*Not*I, RV-*Eco*RV, and H-*Hinf*I

enzymes cut the DNA only if their recognition site is unmethylated. These enzymes cut preferentially in GC-rich promoter regions of genes.

- **Step 4** – The restriction ends are labeled with radioactive nucleotides. A modified version of DNA polymerase, called Sequenase, is used to extend the 4-base overhang and to fill in complementary radioactively labeled nucleotides. Each restriction site is labeled at the end with exactly four nucleotides. Most of the *Not*I restriction sites are located in sequences that are found only once in the genome (single copy). However, some *Not*I sites are located in sequences that appear multiple times in the genome (repetitive sequences). The *Not*I sites from single-copy sequences have single-copy intensity in an RLGS profile, and the *Not*I sites from repetitive sequences are visible as enhanced fragments on the RLGS profile (Fig. 3.3A).

A.

1st-dimension

B.

2nd-dimension

Fig. 3.5 RLGS equipment: **A.** 1st-dimension RLGS gel cylinder; **B.** 2nd-dimension RLGS gel box

- **Step 5** –The DNA is digested with a second restriction enzyme (*Eco*RV) that is not methylation-sensitive.
- **Step 6** – The restriction fragments are separated in a 0.8% agarose gel by electrophoresis. This gel is contained in a Teflon tube and has an inner diameter of 2.4 mm. A single 1st-dimension gel cylinder can hold up to nine 1st-dimension gels (Fig. 3.5A). Both labeled and unlabeled fragments are separated by molecular size in a 60-cm long gel.
- **Step 7** – The agarose gel is expelled from the Teflon tube and transferred into a slightly larger tube. The third restriction enzyme is added at high concentration and the DNA is digested within the gel.
- **Step 8** – The 1st-dimension gel with the digested DNA is transferred to the top of a 2nd-dimension acrylamide gel. After connecting the 1st-dimension gel to the 2nd-dimension gel, the DNA fragments are separated again by molecular size within the acrylamide gel located in a gel box that holds 4 gels (Fig. 3.5B).
- **Step 9** – The acrylamide gels are transferred to filter paper, dried onto the paper using a gel dryer, and subsequently exposed to X-ray film. The X-ray film is developed, and two RLGS images are superimposed and scored for their differences (see Fig. 3.3B for an example).

3.2.3
Applications

In a recent study, Costello and colleagues studied aberrant DNA methylation in multiple human cancers using RLGS [13]. In this global analysis, the methylation status of 1184 CpG islands, represented by RLGS fragments, was tested in 98 primary human tumors. An estimated average of ~600 CpG islands (range, 0–4500) of the 45 000 total CpG islands in the genome were aberrantly methylated in the tumors. This study also showed patterns of CpG island methylation that were common to several types of tumors (methylated in more than one tumor type) and targets that displayed distinct tumor type specificity (methylated in only one tumor type but never in one of the other tumors studied). In addition, the methylation patterns that characterize certain tumors were not random. This means that, either certain sequences become methylated in certain tumors, or methylation occurs at random sites, but selective forces favor the growth of cells with a certain methylation pattern. It is possible to clone RLGS fragments by using a *NotI-Eco*RV library. The majority of *NotI-Eco*RV sequences have been cloned into a plasmid vector and can be multiplied in bacteria. Using RLGS mixing gels enabled certain RLGS fragments to be cross-referenced with the corresponding *NotI-Eco*RV clones in the library. Those clones that represent certain RLGS fragments of interest can be sequenced and subsequently allow the molecular characterization of genes that become methylated in a tumor. In this way, several novel genes that become methylated and silenced by methylation in lung and brain tumors as well as in leukemias have been identified [13–15].

RLGS was also used in two studies to identify allele-specific methylation in the mouse genome. Allele-specific methylation is a rare phenomenon wherein a sequence is methylated on only one allele, either the one inherited from the father or the one inherited from the mother. Allele-specific methylation is associated with genes that show imprinted expression. Imprinted genes are expressed either from the paternal or maternal allele and are responsible for important functions during development. The study is based on the identification of polymorphic RLGS fragments between two inbred mouse strains. These polymorphic fragments have different mobility in RLGS profiles from the two strains. An F1 mouse that inherits both polymorphic alleles will have both RLGS fragments present, but with only half the intensity of the parental strains. Allele-specific methylation in a polymorphic fragment should result in the presence of this fragment in one cross, but its absence in the reciprocal cross. RLGS was the first method that allowed systematic screening for loci that showed allele-specific methylation. The cloning of those sequences that were methylated in an allele-specific pattern gave access to novel imprinted loci. So far, two new imprinted genes (*U2afbprs* and *Grf1*) have been identified in an RLGS scan [16, 17].

3.3

Methylation-sensitive Arbitrarily Primed (AP) PCR: Fishing for the 5th Bases in Genomic Ponds

3.3.1
Principle

The AP-PCR method is based on the ability of PCR to generate a reproducible group of DNA fragments when the reaction is performed at low annealing temperature. Under this condition, a primer may be less stringent in "finding" the exact complementary DNA and hybridize to other, imperfectly matched sequences. As a result of this random association, multiple different DNA fragments can be produced by AP-PCR. To identify methylated sites in the genome, AP-PCR can be performed on DNA samples digested with methylation-sensitive restriction enzymes. As described earlier, methylation-sensitive *Hpa*II and its isoschizomer *Msp*I can be employed in this approach. In addition, a second restriction enzyme *Rsa*I, unrelated to DNA methylation, is used to cut the DNA into smaller fragments, which can reduce potential PCR artifacts that may be generated in the amplification step.

The amplified DNA fragments are separated on the basis of size by gel electrophoresis, displaying a reproducible array of bands. A band is identified as "methylated" if a PCR product is present in both the *Rsa*I-digested and the *Rsa*I + *Hpa*II doubly digested samples, but not in the *Rsa*I + *Msp*I doubly digested sample (Fig. 3.6A). This band is due to the presence of a CCGG site flanked by two primers that are used to amplify a DNA fragment in the *Rsa*I-digested sample. The internal cytosine of this CCGG site is methylated and cannot be digested by *Hpa*II, so only this *Rsa*I fragment is amplified by AP-PCR. In contrast, this site is cut by *Msp*I irrespective of its methylation status and cannot be amplified.

A band is identified as "unmethylated" when a PCR product is present in the *Rsa*I-digested sample only, but not in doubly-digested samples (Fig. 3.6A). Samples digested with *Rsa*I and *Rsa*I + *Msp*I are controls to determine whether the bands observed after the *Rsa*I + *Hpa*II digestion are truly methylated in their internal CCGG sequence. Four possible outcomes – (i) hypomethylated: methylated in adjacent normal tissue and unmethylated in tumor; (ii) hypermethylated: unmethylated in adjacent normal tissue and methylated in tumor; (iii) normally methylated; and (iv) unmethylated – are distinguishable by methylation-sensitive AP-PCR between normal and tumor tissue (Fig. 3.6B).

To identify DNA regions that demonstrate methylation changes in tumors, AP-PCR fragments showing methylation differences are isolated from the gel and cloned into a vector, which is a DNA molecule that can propagate in a bacterial host. The DNA sequences of the cloned fragments are compared with the genome databases to determine whether they match any previously known genomic sequences. If no match is found, the cloned fragments can be used as probes in Northern blotting, a counterpart of the Southern blotting technique, to determine whether these sequences contain new RNA transcripts. Based on the result of Northern analysis, the cloned fragments are then used to screen a complementary DNA library for clon-

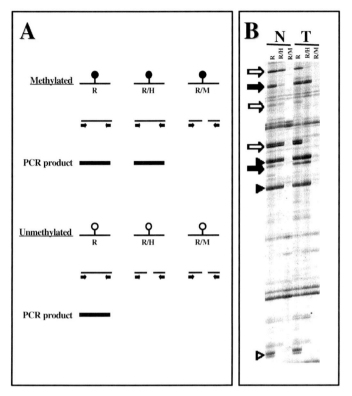

Fig. 3.6 **A.** Schematic outline of the steps in methylation-sensitive AP-PCR. Primers are represented by arrows. Filled lollipops represent methylated CpG sites, and unfilled lollipops represent unmethylated CpG sites. **B.** Examples of methylation-sensitive AP-PCR. Possible outcomes of DNA methylation changes in tumor detected by methylation-sensitive AP-PCR. Solid arrows indicate hypermethylated fragments, open arrows indicate hypomethylated fragments, solid triangles indicate unchanged methylated fragments, and open triangles indicate unchanged unmethylated fragments. N, DNA from normal bladder tissue; T, DNA from bladder tumor cell line (T24); R, *Rsa*I digestion; R/H, *Rsa*I + *Hpa*II digestion; and R/M, *Rsa*I + *Msp*I digestion.

ing novel genes. Thus, methylation-sensitive AP-PCR can be used for identifying methylation changes at CpG islands and for isolating new genes associated with this alteration in cancer.

3.3.2
How Does MS AP-PCR Work?

- **Step 1** – Genomic DNA is digested with *Rsa*I and methylation-sensitive *Hpa*II or its isoschizomer *Msp*I. To identify additional regions of DNA associated with altered methylation, other methylation-sensitive restriction enzymes (*Mvn*I, *Ksp*I, and *Hha*I) can be used in this step.

- **Step 2** – Restriction-digested DNA (100–200 ng) is amplified in the presence of radiolabeled tracer (α^{32}P-dCTP) using AP-PCR with a single primer or a combination of 2 or 3 primers. The methylation fingerprints generated by AP-PCR are highly dependent on the sequences of the arbitrary primers. Our studies showed that GC-rich primers are more successful at amplifying GC-rich DNA fragments, because there is a high probability that they preferentially anneal to sequences associated with CpG islands. We have designed primers between 10 and 20 bases in length. For 20-base primers, a GC rich sequence is used at the 3′ end. For 10-base primers, a very high GC content (80%–100%) is used. Primer length also affects the number of bands, with shorter arbitrary primers yielding better results.
- **Step 3** – PCR products are separated in 5% polyacrylamide gels. Radiolabeled DNA fragments are identified after exposure to X-ray film. Candidate bands that are differentially methylated are excised from polyacrylamide gels and eluted in sterile water. The eluate is then used in a PCR reaction with the same primer(s) used in the original AP-PCR to generate sufficient amounts of template for vector cloning and DNA sequencing.
- **Step 4** – The resulting nucleotide sequences are then compared with the GenBank sequences, using the BLAST program (http://www.ncbi.nlm.nih.gov/BLAST/). Often, Southern blotting analysis is necessary to confirm that the cloned band corresponds to the band visualized by AP-PCR.

3.3.3
Applications

In a recent study, Liang and colleagues used methylation-sensitive AP-PCR to detect and isolate many regions of genomic DNA that had undergone methylation changes during tumorigenesis [18, 19]. Fig. 3.7 shows a detailed example of a DNA band con-

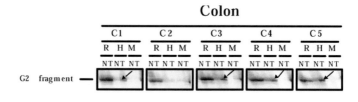

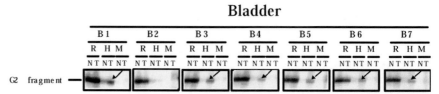

Fig. 3.7 Details of typical methylation-sensitive AP-PCR gels, showing bands (indicated by arrows) of a hypermethylated fragment, G2, present in colorectal and bladder tumors (see text for explanation). N, DNA from adjacent normal tissue; T, tumor DNA; R, *Rsa*I digestion; H, *Rsa*I + *Hpa*II digestion; M, *Rsa*I + *Msp*I digestion.

taining hypermethylated sites in four of five (80%) colon tumors and six of seven (85%) bladder tumors compared with their paired normal controls. As of today, ~100 novel CpG islands that are frequently hypermethylated in bladder tumors have been isolated. Using this approach, several novel genes such as *TPEF*, *PAX6*, and *endothelin receptor B* have been shown to be differentially methylated in many cancers [20–22].

In addition to isolating unknown sequences associated with methylation changes in genomic DNA, methylation-sensitive AP-PCR was used to rapidly quantitatively estimate the variability in methylation at multiple sites between different cell lines or between normal and tumor tissues [18, 19, 23]. White blood cells (WBCs), tumor tissue (colon, bladder, and prostate cancer) tissue, and adjacent normal tissue from 17 patients were screened by methylation-specific AP-PCR [19]. DNA methylation pattern analysis based on 45 positive bands showed little interindividual differences in WBCs and adjacent normal tissue samples, but showed some tissue-specific differences. However, cancer cells showed marked methylation changes that varied considerably between different tumors, suggesting different methylation patterns in patients. Based on the number of putative methylation sites detected by methylation-sensitive AP-PCR in matched sets of normal and tumor DNAs, the major differences observed were bands representing putative regions of hypo- or hypermethylation in tumors relative to normal samples. Regardless of the primer combination used for methylation-sensitive AP-PCR, hypomethylated regions of DNA were consistently associated with CpG-poor sequences, whereas hypermethylated regions of DNA correlated with CpG-rich sequences. These findings are consistent with the types of methylation changes associated with tumorigenesis [24]. These studies demonstrated that methylation-sensitive AP-PCR can therefore be used to identify new CpG islands that may become differentially methylated during tumorigenesis.

3.4
Differential Methylation Hybridization (DMH): Identifying the 5th Bases in the Genomic Crossword Puzzle

3.4.1
Principle

DMH is an array-based approach for screening methylation changes of CpG islands in the genome [25, 26]. The array consists of many short CpG island fragments spotted at specific locations on an "affinity matrix" coated on a glass slide surface. This type of nucleic acid array (often called a DNA microarray or chip) can contain more than 5000 DNAs per square centimeter, dramatically increasing the experimental efficiency and information content. The microarray hybridization works in a manner similar to northern or Southern blotting. In this new approach, the immobilized nucleic acid, arrayed on the surface, is called the "probe", and the free nucleic acid in solution used for array hybridization is called the "target." The hybridization is, in effect, a highly parallel search by each target for a matching probe attached to the glass surface. The bound DNA molecules, which emit fluorescence signals, are then detected by a microarray scanner.

To prepare a DMH target, DNA is digested with a 4-base restriction enzyme known to cut DNA frequently into small fragments (< 200 base pairs) but to preserve larger GC-rich CpG island fragments relatively intact (Fig. 3.8). The ends of these GC-rich fragments are ligated to linkers and then restricted by methylation-sensitive enzymes. When an internal recognition sequence(s) of a DNA fragment is abnormally methylated in a tumor genome, it cannot not be digested by the methylation-sensitive enzyme, and the DNA fragment is amplified by PCR with the flanking primers. In contrast, this DNA fragment is unmethylated in the normal genome and can be digested away by the enzyme and thus not be amplified in normal control DNA. After linker-PCR, many amplified fragments are produced, creating a different pool of the DNA population in tumor relative to normal control, due to abnormal DNA methylation in the tumor genome.

The amplified tumor and normal DNAs are labeled with fluorescent dyes Cy5 and Cy3, respectively, and mixed in solution for microarray hybridization. Tumor and normal targets compete for binding to the arrayed CpG island probes. When equal amounts of tumor and normal targets bind to a probe, the bound DNA emits equal red and green fluorescence signals detected by a scanner. The merged signals produce a yellow spot, indicating equal presence of the same target in both tumor and control samples that are not digested away by the methylation-sensitive enzyme (Fig. 3.9). This result suggests that the corresponding CpG island is likely methylated in both tumor and normal genomes. A probe hybridized predominantly with tumor DNA, but not with normal DNA, appears red and indicates hypermethylated CpG islands present in the tumor genome. A green spot denotes hypomethylation of a normally methylated CpG island in tumor DNA.

3.4.2
How Does DMH Work?

DMH consists of three main components: preparation of the CpG island microarray, generation of the fluorescently labeled targets, and analysis of microarray data.

- **Step 1** – A genomic library, CGI, prepared to contain thousands of cloned CpG island fragments was obtained from the Human Genome Mapping Program Cen-

---

Fig. 3.8 Schematic flowchart for differential methylation hybridization. DNA samples are first digested with *MseI*, a four-base cutter that restricts bulk DNA into smaller fragments (< 200 bp) but leaves the GC-rich CpG islands relatively intact. These GC-rich fragments are next ligated to linkers, restricted with methylation-sensitive endonucleases, and amplified by PCR from the linkers. Methylated DNA fragments in the tumor sample are protected from restriction and amplified, and unmethylated fragments in the normal sample are cut and cannot be amplified. These amplified products thus contain different pools of DNA fragments that reflect methylation differences between tumor and normal DNAs. Tumor and normal DNA are subsequently labeled with Cy5 and Cy3 dyes, respectively, and cohybridized on a microarray chip containing 8000 CpG-island tags.

Differential Methylation Hybridization

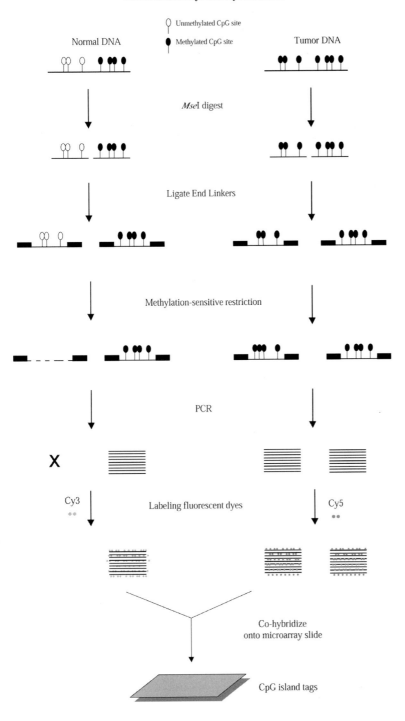

CpG Island Microarray

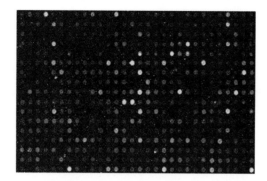

Yellow spots: No methylation changes
Red spots: Hypermethylation
Green spots: Hypomethylation

Fig. 3.9 Representative results of differential methylation hybridization. For each experiment, the tumor amplicon is labeled with the Cy5 (red) fluorescent dye and cohybridized onto the CpG-island microarray with a normal control amplicon labeled with Cy3 (green) dye. The output is the measured intensities of Cy5 and Cy3 fluorescence and indicates the amount of bound DNA for a CpG island. Yellow spots indicate equal amounts of methylated DNA in the tumor and control amplicon, indicating no methylation differences between their genomes. Red spots signify a greater presence of bound DNA in tumor relative to normal amplicon, indicating DNA hypermethylation in the tumor genome. Green spots denote loss of methylation (or hypomethylation) from normally methylated DNA.

ter, United Kingdom. Approximately 8000 individual CGI bacterial clones were organized in 96-well culture plates.

- **Step 2** – A 96-pin replicator is used to simultaneously transfer colonies from culture plates to 96-well PCR tubes. PCR is conducted to amplify CpG island fragments using primers derived from the vector sequences.
- **Step 3** – A robotic device, the Affymetrix/GMS 417 Arrayer, is used to deposit ~0.05 µl of each PCR product onto a polylysine-coated surface that covalently binds DNA to the glass slides. A total of 8000 CpG island fragments can be dotted onto a 4.5 × 1.6 cm area of a microscope slide. After postprocessing to remove unbound DNA, the microarray slides are ready for hybridization.
- **Step 4** – To prepare targets, DNA from test and control samples is digested with *Mse*I, whose recognition site TTAA rarely occurs in GC-rich regions. The enzyme cuts genomic DNA into very small fragments but leaves most CpG islands intact. The digested ends of CpG island fragments are ligated to linkers, which are used for anchoring sequences of a PCR primer.
- **Step 5** – The ligated DNA is digested with methylation-sensitive enzymes *Hpa*II and *Bst*UI.
- **Step 6** – The digested DNA is amplified by PCR using a primer that binds to the flanking linkers. DNA fragments containing methylated sites cannot be digested

by the methylation-sensitive enzymes and are amplified by this linker-PCR approach. Many of these amplified fragments are expected be present in tumor, due to abnormal DNA methylation, whereas the same unmethylated fragments are digested and not present in the normal control.

- **Step 7** – Amplified target DNA is purified, and aminoallyl-dUTP is incorporated into DNA through a random primed labeling procedure. The Cy5 or Cy3 fluorescent dye is then coupled to the aminoallyl-dUTP residue in the incorporated site.
- **Step 8** – The fluorescently labeled tumor and normal targets are pooled and hybridized to a microarray slide. Blocking reagents are included to minimize hybridization noise.
- **Step 9** – Hybridized slides are scanned with a GenePix 4000A scanner. Laser excitation of the hybridized targets bound to probes yields an emission pattern that is collected and exported as a numerical .gpr file. The digital image is stored as a .tif file.
- **Step 10** – Each DMH experiment is expected to generate thousands of data points. Specialized bioinformatics tools have been developed to analyze large data sets. One such program is hierarchical clustering, which can be used to identify different patterns of DNA methylation in tumors and to find specific epigenetic "fingerprints" that are associated with patients' clinical features.

3.4.3
Applications

DMH was used to query the methylation status of ~8000 CpG islands in a panel of 17 paired tissues of breast tumors and normal controls [26]. Microarray data analysis showed that a wide spectrum of DNA hypermethylation, ranging from 15 to 207 CpG islands, was observed in this patient group. Hierarchical clustering, which identified the degrees of similarity in DNA methylation among these tumors, identified 3 clusters of breast tumors. Two of these clusters consisted of tumors deficient in estrogen and progesterone receptors; such tumors are less responsive to chemotherapy. The third group, showing less DNA methylation, was made up of tumors that expressed these hormone receptors and had a better prognosis. In a second study, DMH was performed on 19 ovarian tumors [27]. The total number of hypermethylated CpG islands ranged from 48 to 348 in this ovarian tumor group. Hierarchical clustering further separated these tumors into 2 groups based on their methylation profiles. The first group had lower methylation levels and was largely composed of patients that responded to chemotherapy. The second group of patients showed extensive methylation of CpG islands and had poorer survival after chemotherapy. From these two studies, important CpG islands that were more susceptible to methylation alteration were identified and have been used as landmarks to clone associated genes whose expression is subject to methylation control in cancer cells. In addition, these CpG islands are potentially useful as epigenetic markers for cancer diagnosis and prognosis.

3.5
MethyLight: Finding 5th-base Patterns in Genomic Shadows

3.5.1
Principle

The principal difference between MethyLight and the technologies discussed above is illustrated in Fig. 3.10. RLGS, AP-PCR, and DMH analyze multiple CpG dinucleotides scattered throughout the genome. Each of these CpGs is analyzed collectively for all the DNA molecules present in the sample (vertical columns), and each CpG in the genome (horizontal axis) is analyzed independently of the other CpGs. The result is a profile of 5th bases in the genome. In contrast, MethyLight looks for specific patterns of 5th bases in individual DNA molecules (horizontal lines). The most common pattern analyzed is full methylation of 5 to 10 closely linked CpGs. Such fully methylated patterns can be relatively uncommon (Fig. 3.10). A unique feature of MethyLight is its ability to detect such rare methylation patterns in the presence of a vast excess of other methylation patterns. MethyLight can detect fully methylated molecules on a background of a 10000-fold excess of unmethylated molecules [28].

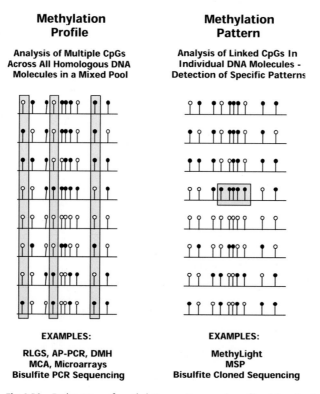

Methylation Profile

Analysis of Multiple CpGs Across All Homologous DNA Molecules in a Mixed Pool

Methylation Pattern

Analysis of Linked CpGs In Individual DNA Molecules - Detection of Specific Patterns

EXAMPLES:

RLGS, AP-PCR, DMH
MCA, Microarrays
Bisulfite PCR Sequencing

EXAMPLES:

MethyLight
MSP
Bisulfite Cloned Sequencing

Fig. 3.10 Explanation of methylation pattern and profile determined by various genomic approaches (see text for explanation)

The MethyLight technique relies on the bisulfite modification of genomic DNA to create methylation-dependent sequence changes (see Sect. 3.1.2). PCR primers can be designed to amplify only specific variants of such sequence patterns (such as fully methylated or fully unmethylated versions). This principle is called Methylation-Specific PCR, or MSP [10]. In conventional MSP, the PCR products are detected by agarose gel electrophoresis. In MethyLight, detection of the PCR products occurs in real time during the PCR reaction, using fluorescence.

One variant of such real-time PCR is called TaqMan® technology, which includes a third oligonucleotide in the PCR reaction, which anneals in between the two amplification primers as a probe (Fig. 3.11). The probe has a fluorescent group (F) covalently attached to the 5′ end of the oligonucleotide. However, the fluorescence of this labeled molecule is suppressed by the inclusion of a quencher (Q) covalently attached to the 3′ end of the probe. The first step in a PCR cycle is denaturation of the double-stranded DNA. Subsequently, the temperature is lowered to allow annealing of the primers. At this step, the probe also anneals, downstream of the amplification primer (Fig. 3.11).

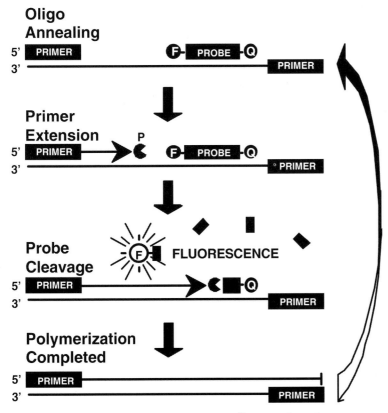

Fig. 3.11 Schematic outline of the steps in TaqMan® (see text for explanation). P: DNA polymerase; F: fluorescent tag; and Q: quencher

Then the polymerase (P) extends the primer to synthesize the new DNA strand. When the polymerase reaches the probe, the 5′-to-3′ exonuclease activity of the polymerase degrades the probe. This cleavage step liberates the fluorescent label from the proximity of the 3′ quencher, allowing fluorescence. The fluorescence is evoked by illuminating the PCR reaction mixture at the correct excitation wavelength, usually with a halogen or argon laser. The level of fluorescence for the PCR reaction can then be measured. The amount of a particular genomic DNA sequence initially present in the reaction can be derived from the fluorescence measured during the course of the PCR reaction. By designing primers and probes that correspond to the bisulfite-modified version of a particular methylation pattern in a region of interest, the investigator can rapidly and sensitively assess the relative amounts of this methylation pattern in the DNA sample.

3.5.2
How Does MethyLight Work?

- **Step 1** – Genomic DNA is isolated from a tissue sample. The DNA does not need to be of high quality. The MethyLight technique is even compatible with fragmented genomic DNA isolated from paraffin-embedded, formalin-fixed tissue samples.
- **Step 2** – The genomic DNA is denatured at high temperature and high pH.
- **Step 3** – The DNA is then incubated in the presence of high concentrations of sodium bisulfite. During this incubation, which can last from several hours to overnight, the cytosine residues are sulfonated and undergo deamination (Fig. 3.2).
- **Step 4** – After the bisulfite incubation, the cytosine bases, which have now been converted to uracil residues, are desulfonated at high pH.
- **Step 5** – The DNA sample is then neutralized and purified for storage and analysis.
- **Step 6** – PCR reactions are set up with the bisulfite-modified DNA. Each reaction includes a fluorescent probe, in addition to the forward and reverse PCR primers (Fig. 3.11). MethyLight can be performed in individual tubes, 96-well plates, or 384-well plates to allow for higher throughput.
- **Step 7** – The real-time PCR reactions are run in dedicated instruments, which output data on the relative amounts of measured DNA template in each tube or well in a spreadsheet format.
- **Step 8** – The data are then processed for further statistical analysis.

3.5.3
Applications

The MethyLight technology has three main strengths. One is its ability to detect rare patterns of DNA methylation in a sample with a preponderance of other patterns. The second is its compatibility with low-quality genomic DNA, such as that derived from paraffin-embedded, formalin-fixed tissue samples. Its third strength is its ability to rapidly screen for several different genes in a fairly large number of samples.

Since it does not rely on the presence of restriction-enzyme sites, MethyLight reactions can be designed for virtually any gene or CpG island of interest. Therefore, specific panels of methylation reactions can be developed and tested on tissue samples. An example of this approach is the analysis of the methylation status of 20 different gene loci in a study of 84 tissue samples derived from patients with esophageal adenocarcinoma [29]. This study revealed that different loci become abnormally methylated at various steps in the progression of the cancer.

MethyLight technology can be applied to the sensitive detection of abnormally methylated tumor DNA in cancer patients. The technique is sufficiently sensitive to detect low amounts of tumor DNA in the circulating blood of cancer patients. For instance, in a study of 52 patients with esophageal adenocarcinoma, MethyLight could detect abnormally methylated DNA in 25% of the patients, whereas none of 54 control individuals showed detectable levels of abnormally methylated DNA [30]. We anticipate that this ability of MethyLight to detect low quantities of abnormally methylated tumor DNA may lead to new techniques for the early detection of cancer.

3.6
Exploring the Epigenome

The complexity of DNA methylation information in the genome appears to be almost matched by the diversity of methods used to extract this information. This chapter presents an overview of some of the techniques used in DNA methylation research. However, many other useful methods are not covered here. This chapter focuses on techniques used in a genomic approach to understanding DNA *methylation patterns* and *profiles* (Fig. 3.10).

Two simpler measures of DNA methylation are *methylation content* and *methylation level*. Methylation content, or 5-methylcytosine content, refers to the total amount of 5-methylcytosine present in a DNA sample. This measurement of base composition does not provide any information on how the 5th bases are distributed, but it is a useful genome-wide measurement that can complement the higher-resolution methylation profiles obtained with the techniques described in this chapter.

Methylation *level* refers to the average amount of methylation present at an individual CpG dinucleotide. The example of a methylation *profile* shown in Fig. 3.10 consists of measurements of methylation *levels* at three different CpG dinucleotides. Techniques that can measure methylation levels with much greater accuracy than the profiling techniques described in this chapter have been developed, but they are limited in scope, since they are labor-intensive and assess methylation at just one CpG at a time. They are therefore more suited for targeted methylation analyses.

The plethora of 5th-base techniques necessitates careful evaluation of the available options to ensure choice of the appropriate tool. Exploration and exploitation of the epigenome can be divided into four stages: *discovery, evaluation, validation,* and *implementation*. The *discovery* phase refers to the identification of differentially methylated sites in the genome that could potentially provide molecular diagnostic information. These novel sites are subsequently evaluated in a large number of samples to assess their utility. This *evaluation* step allows a small number of highly informative sites to be selected from the large number initially subjected to evaluation. This process leads to the identification of a relatively small panel of markers with potential molecular diagnostic value. This panel is then subjected to a *validation* step, during which the performance of the panel is assessed on a large number of unknown samples and compared with standard diagnostic methods for accuracy. The successful validation of a panel finally leads to its *implementation* as a routine diagnostic method.

Each of the techniques described in this chapter is particularly useful at different phases in exploration and exploitation of the epigenome. RLGS is very efficient at both the discovery of new candidate sites and their evaluation, with a moderate number of samples. It is less well suited for validation and implementation, since it does not provide a way to select a small number of highly informative sites for screening a large number of samples.

MS AP-PCR is very useful for the discovery of differentially methylated sites. It is not as efficient as RLGS in screening large numbers of sites simultaneously, but it is easier to implement, and it provides a more direct method of identifying the differentially methylated sites.

DMH is more flexible than the previous two techniques, since the investigator can adjust the composition of the CpG island microarray arranged on the chip. In the discovery phase, 8000 individual CGI PCR products are arranged on the chip. However, a more specialized chip containing a small subset of the 8000 sites could be constructed for panel validation, if desired. Implementation of a diagnostic panel using DMH may be problematic if the genomic DNA of routine samples is of low quality, such as in formalin-fixed tissues.

MethyLight can be used for the assessment of specific methylation patterns at multiple loci. This application of MethyLight yields a profile of methylation patterns, rather than a profile of individual CpGs. Nevertheless, it is much less efficient at such parallel analyses than the three restriction-enzyme-based methods described above. MethyLight does excel in parallel processing of a large number of samples,

MethyLight: Automated DNA Methylation Analysis

Fig. 3.12 The automated MethyLight system for high-throughput DNA methylation analysis

since it is easily scalable and is amenable to automated setup and analysis. A typical setup of a 96-well based, automated MethyLight system is shown in Fig. 3.12. These features make MethyLight well suited for validation and implementation of relatively small panels of diagnostic markers.

Now is an exciting period in the field of epigenomics. New tools have provided the first access to the complex information stored in the epigenome. The race is now on to discover and exploit this rich source of molecular information. Much of the initial focus has been on cancer molecular diagnostics. However, as the field develops and broadens, we can expect to see applications in many other areas of human health research, as well as in forensics, agriculture, and other fields.

References

1 CHRISTMAN, J. K. (**1982**) *Anal. Biochem.* 119, 38–48.

2 WANG, R. Y.; GEHRKE, C. W.; EHRLICH, M. (**1980**) *Nucleic Acids Res.* 8, 4777–4790.

3 ESTELLER, M.; CORN, P. G.; BAYLIN, S. B.; HERMAN, J. G. (**2001**) *Cancer Res.* 61, 3225–3229.

4 BIRD, A. P. (**1986**) *Nature* 321, 209–213.

5 JONES, P. A.; LAIRD, P. W. (**1999**) *Nature Genet.* 21, 163–167.

6 BAYLIN, S. B.; ESTELLER, M.; ROUNTREE, M. R.; BACHMAN, K. E.; SCHUEBEL, K.; HERMAN, J. G. (**2001**) *Hum. Mol. Genet.* 10, 687–692.

7 EHRLICH, M. (**2000**), in *DNA Alterations in Cancer*, ed. Ehrlich, M., Eaton Publishing, Natick, MA, 273–291.

8 TOYOTA, M.; HO, C.; AHUJA, N.; JAIR, K. W.; LI, Q.; OHE-TOYOTA, M.; BAYLIN, S. B.; ISSA, J. P. (**1999**) *Cancer Res.* 59, 2307–2312.

9 USHIJIMA, T.; MORIMURA, K.; HOSOYA, Y.; OKONOGI, H.; TATEMATSU, M.; SUGIMURA, T.; NAGAO, M. (**1997**) *Proc. Natl. Acad. Sci. USA* 94, 2284–2289.

10 HERMAN, J. G.; GRAFF, J. R.; MYOHANEN, S.; NELKIN, B.; BAYLIN, S. B. (**1996**) *Proc. Natl. Acad. Sci. USA* 93, 9821–9826.

11 GONZALGO, M. L.; JONES, P. A. (**1997**) *Nucleic Acids Res.* 25, 2529–2531.

12 XIONG, Z.; LAIRD, P. W. (**1997**) *Nucleic Acids Res.* 25, 2532–2534.

13 COSTELLO, J. F.; FRUHWALD, M. C.; SMIRAGLIA, D. J.; RUSH, L. J.; ROBERTSON, G. P.; GAO, X.; WRIGHT, F. A.; FERAMISCO, J. D.; PELTOMAKI, P.; LANG, J. C.; SCHULLER, D. E.; YU, L.; BLOOMFIELD, C. D.; CALIGIURI, M. A.; YATES, A.; NISHIKAWA, R.; SU HUANG, H.; PETRELLI, N. J.; ZHANG, X.; O'DORISIO, M. S.; HELD, W. A.; CAVENEE, W. K.; PLASS, C. (**2000**) *Nature Genet.* 24, 132–138.

14 DAI, Z.; LAKSHMANAN, R. W. Z.; SMIRAGLIA, D. J.; RUSH, L. J.; FRUEHWALD, M.; BRENA, M.; LI, B.; WRIGHT, F. A.; ROSS, P.; OTTERSON, G. A.; PLASS, C. (**2001**) *Neoplasia* 3, 314–323.

15 RUSH, L. J.; DAI, Z.; SMIRAGLIA, D. J.; GAO, X.; WRIGHT, F. A.; FRUHWALD, M.; COSTELLO, J. F.; HELD, W. A.; YU, L.; KRAHE, R.; KOLITZ, J. E.; BLOOMFIELD, C. D.; CALIGIURI, M. A.; PLASS, C. (**2001**) *Blood* 97, 3226–3233.

16 PLASS, C.; SHIBATA, H.; KALCHEVA, I.; MULLINS, L.; KOTELEVTSEVA, N.; MULLINS, J. I.; KATO, R.; SASAKI, H.; HIROTSUNE, S.; OKAZAKI, Y.; HAYASHIZAKI, Y.; CHAPMAN, V. M. (**1996**) *Nature Genet.* 14, 106–109.

17 HAYASHIZAKI, Y.; SHIBATA, H.; HIROTSUNE, S.; SUGINO, H.; OKAZAKI, Y.; SASAKI, N.; HIROSE, K.; IMOTO, H.; OKUIZUMI, H.; MURAMATSU, M. (**1994**) *Nature Genet.* 6: 33–40.

18 GONZALGO, M. L.; LIANG, G.; SPRUCK, C. H.; ZINGG, J.-M.; RIDEOUT, W. M.; JONES, P. A. (**1997**) *Cancer Res.* 57, 594–599.

19 LIANG, G.; SALEM, C. E.; YU, M. C.; NGUYEN, H. D.; GONZALES, F. A.; NICHOLS, P. W.; JONES, P. A. (**1998**) *Genomics* 53, 260–268.

20 LIANG, G.; ROBERTSON, K. D.; TAL-
MADGE, C.; SUMEGI, J.; JONES, P. A.
(2000) *Cancer Res.* 60, 4907–4912.

21 SALEM, C. E.; MARKL, I. D.; BENDER,
C. M.; GONZALES, F. A.; JONES, P. A.;
LIANG, G. (2000) *Intl. J. Cancer* 87, 179–
185.

22 PAO, M. M.; TSUTSUMI, M.; LIANG, G.;
UZVOLGYI, E.; GONZALES, F. A.; JONES,
P. A. (2001) *Hum. Mol. Genet.* 10, 903–
910.

23 MARKL, I. D.; CHENG, J.; LIANG, G.;
SHIBATA, D.; LAIRD, P. W.; JONES, P. A.
(2001) *Cancer Res.* 61, 5875–5884.

24 LAIRD, P. W.; JAENISCH, R. (1994) *Hum.
Mol. Genet.* 3, 1487–1495.

25 HUANG, T. H.-M.; PERRY, M. R.; LAUX,
D. E. (1999) *Hum. Mol. Genet.* 8, 459–
470.

26 YAN, P. S.; CHEN, C.-M.; SHI, H.; RAH-
MATPANAH, F.; WEI, S. H.; CALDWELL,
C. W.; HUANG, T. H.-M. (2001) *Cancer
Res.* 61, 8375–8380.

27 WEI, S. H.; CHEN, C.-M.; STRATHDEE,
G.; HARNSOMBURANA, J.; SHYU, C.-R.;
RAHMATPANAH, F.; SHI, H.; NG, S. W.;
YAN, P. S.; NEPHEW, K. P.; BROWN, B.;
HUANG, T. H.-M. (2002) *Clin. Cancer
Res.* 8, 2246–2252.

28 EADS, C. A.; DANENBERG, K. D.; KAWA-
KAMI, K.; SALTZ, L. B.; BLAKE, C.; SHI-
BATA, D.; DANENBERG, P. V.; LAIRD,
P. W. (2000) *Nucleic Acids Res.* 28, E32.

29 EADS, C. A.; LORD, R. V.; WICKRAMA-
SINGHE, K.; LONG, T. I.; KURUMBOOR,
S. K.; BERNSTEIN, L.; PETERS, J. H.; DE-
MEESTER, S. R.; DEMEESTER, T. R.; SKIN-
NER, K. A.; LAIRD, P. W. (2001) *Cancer
Res.* 61, 3410–3418.

30 KAWAKAMI, K.; BRABENDER, J.; LORD,
R. V.; GROSHEN, S.; GREENWALD, B. D.;
KRASNA, M. J.; YIN, J.; FLEISHER, A. S.;
ABRAHAM, J. M.; BEER, D. G.; SI-
DRANSKY, D.; HUSS, H. T.; DEMEESTER,
T. R.; EADS, C.; LAIRD, P. W.; ILSON,
D. H.; KELSEN, D. P.; HARPOLE, D.;
MOORE, M. B.; DANENBERG, K. D.; DA-
NENBERG, P. V.; MELTZER, S. J. (2000)
J. Natl. Cancer Inst. 92, 1805–1811.

4
Mammalian Epigenomics: Reprogramming the Genome for Development and Therapy

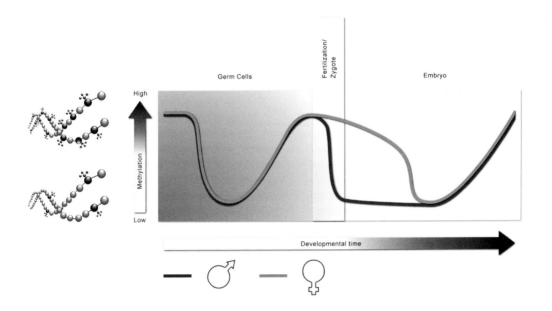

4

Mammalian Epigenomics: Reprogramming the Genome for Development and Therapy

Wolf Reik and Wendy Dean

Summary

Epigenetic modifications of DNA and chromatin are important for genome function during development and in adults. DNA and chromatin modifications have central importance for genomic imprinting and other aspects of epigenetic control of gene expression. In somatic lineages, modifications are generally stably maintained and are characteristic of different specialized tissues. The mammalian genome undergoes major reprogramming of modification patterns in the germ cells and in the early embryo. Some of the factors that are involved both in maintenance and in reprogramming, such as methyltransferases, are being identified. Epigenetic reprogramming is deficient during animal cloning, which is a major explanation for the inefficiency of the cloning procedure. Deficiencies in reprogramming are likely to underlie the occurrence of epimutations and of epigenetic inheritance. Environmental factors can alter epigenetic modifications and may thus have lasting effects on the phenotype. Epigenomics methods are being developed to catalogue genome modifications under normal and pathological conditions. Epigenetic engineering is likely to play an important role in medicine in the future.

4.1
Introduction

Epigenetics is now a rapidly developing, exciting field in plant, animal, and microbial biology. International Gordon and Keystone conferences cover the epigenetics theme, and last year a special issue of *Science* was published to coincide with the 2001 Gordon Conference on epigenetics [1]. The excitement in epigenetics has reached a new level now with the full sequencing of many of the genomes of model organisms. The next level of global understanding will come from systematically cataloguing epigenetic modifications in different tissues and developmental and pathological situations. This has been recognized by coining an apt term for this new area, epigenomics.

Here we briefly summarize several exciting developments in our own area of mammalian epigenetics. We consider DNA methylation and its likely interaction with histone modifications such as acetylation and methylation. We describe some properties of the mechanisms of imprinting, which is one of the best-studied systems of epigenetic gene regulation. We look at recent exciting work on epigenetic reprogramming, and particularly, how reprogramming is relevant for cloning. We examine the occurrence of epimutations and epigenetic inheritance. Finally, we briefly consider how epigenomics techniques will revolutionize the study and applications of epigenetics in medicine.

4.2
DNA Methylation

Methylation of DNA at cytosine-guanine (CpG) dinucleotides exists in virtually all vertebrates, many invertebrates, and many plant species (although these also methylate other nucleotides). Even *Drosophila*, which for many years was thought not to have methylation, has now been shown to have some, although its biological relevance remains to be established [2]. In mammals, methylation patterns can be stably maintained by the enzyme DNA methyltransferase (Dnmt1), which remethylates new DNA strands upon replication if the mother strand was methylated [3, 4]. This property is important for the maintenance of clonal patterns of methylation which are needed, for example, in imprinting. DNA devoid of methylation can be methylated de novo by the enzymes Dnmt3 a and b and perhaps others [5]. Demethylation of DNA can occur when replication happens in the absence of Dnmt1 (passive demethylation). Active demethylation may occur in vivo in early mammalian embryos and in tumor cell lines, but the enzymes have not been identified [6–9]. How specific methylation patterns are imposed in the genome is largely unknown, since the enzymes have little specificity for DNA sequences (but see below for interactions with chromatin).

Everyone agrees that methylation serves several important biological purposes, but there is much disagreement when it comes to defining in more detail some of these purposes. DNA methylation clearly has an important role in imprinting, both in silencing certain genes (e. g., *H19*) as well as in activating others (*Igf2*, *Igf2r*) [10, 11]. X-chromosome inactivation also depends on methylation [12], both for regulation of the *Xist* gene and for maintenance of inactivation. Less well defined is the role of methylation in genome stability, in particular in cancer [13, 14]. But, most surprisingly, the postulated role of methylation in developmental gene regulation as originally envisaged [15, 16] is still debated. Although it is clear that mouse embryos that lack methylation because of a deficiency in Dnmt1 or Dnmt3 a,b die soon after implantation, the cause of death is not known. Altered imprinting and X-chromosome inactivation have been implicated, and more recently, apoptosis [17]. However, it is still unclear whether and how the deficiency in methylation leads to aberrant regulation of a substantial proportion of endogenous genes [4, 18–20]. It has been proposed that methylation is primarily a host defense against intragenomic parasites such as viruses and retro-elements, and some of these are indeed inactivated by methylation [21]. Other functions of methylation (imprinting) would then evolve from the host defense system [22]. However, one clear role of gene repression by methylation is in cancer, where tumors are often globally hypomethylated but locally hypermethylated, especially in tumor-suppressor genes [23–25].

The repressive effects of promoter methylation on gene repression are mediated chiefly through chromatin condensation. A family of proteins binding to methylated DNA (MBDs) has been identified, and some of these are in complexes with histone deacetylases [4]. This leads to local deacetylation of histone tails in methylated DNA and thus a more condensed chromatin, which is inaccessible to the transcriptional machinery. Contrary to expectation, in mouse knockouts of some of these MBDs, effects on the transcription of endogenous genes have not yet been found, although transfected methylated genes can be expressed in knockout cells [4]. It is also surprising that two of these knockouts *(MeCP2, MBD2)*, as well as the human disease associated with mutations in *MeCP2*, Rett syndrome, do not have defects consistent with their presumed functions but have very specific phenotypic effects exclusively in the postnatal central nervous system. It will be interesting to see whether specific target genes of methylation-mediated repression can eventually be identified.

4.3
Histone Modifications

It has long been known that histone proteins that form the nucleosome are modified by various chemical additions, including phosphorylation, acetylation, and methylation. However, researchers have only recently realized that these modifications are dynamic during development, vary among different tissues, are regulated by specific enzymes, play major roles in the control of gene expression, and interact with other epigenetic control systems such as DNA methylation [26–28]. Acetylation of various amino acid residues of histones H3 and H4 is generally associated with an active chromatin configuration and expressed genes (euchromatin). In contrast, histone methylation is generally associated with condensed or heterochromatic chromatin and gene repression. However, many exceptions to this simple rule exist, and the interactions are highly dynamic and complex. One of the most exciting findings of recent years is that these epigenetic systems interact with each other (thus resulting in the so-called histone code of developmental gene expression), as well as with the DNA methylation system [29]. For example, an H3 lysine-9 acetyl group (potentially associated with gene expression) that is removed by histone deactelyase (HDAC) can then become methylated at the same position by histone methyltransferases (such as the ones known as Suvars (suppression of variegation in *Drosophila*)). This can create a binding site for heterochromatin protein 1 (HP1) or other repressive chromatin factors. HP1 can result in binding of a chromomethylase (at least in *Arabidopsis*), which in turn leads to DNA methylation and stable silencing [30]. Presumably, the epigenetic cycle of repression is closed by binding MBDs and by the recruitment of HDACs and (more speculatively) histone methyltransferases [4, 26–28]. How the cycle of repression is broken to allow gene expression is less clear, since neither active demethylases nor histone demethylases are known. Currently, one has to assume that maintenance methylation of DNA and of histones is inhibited, while acetylases are activated, possibly by transcription itself.

These important new insights result in great advances in our thinking about how patterns of DNA methylation might be created, maintained, and perhaps erased. Further exciting developments in this area are expected from studies in many different species.

4.4
Imprinting

A proportion of genes in mammals (estimated at 0.1%–1% of all genes) and in flowering plants are repressed on one of a pair of chromosomes, and this is depends on the parental origin of the gene. This is called genomic imprinting and explains why in mammals, parthenogenetic development from eggs only is not possible [10, 11, 31]. In mammals, imprinted genes are especially implicated in the regulation of fetal growth, development and function of the placenta, and postnatal behaviors. In fetal growth regulation, imprinted gene action exhibits a peculiar directionality, with the majority of paternally expressed genes enhancing fetal growth, but the majority of maternally expressed genes suppressing fetal growth. This pattern is explained by the genetic-conflict theory of imprinting, which proposes that paternal alleles for genes that increase fetal size by extracting more nutrients from the mother should be selected to be as greedy as possible. Maternal alleles, in contract, need to be more conservative, since the total reproductive success of a female could be compromised by giving individual offspring too many resources [32]. This leads, so the theory says, to imprinted expression of these particular genes. No satisfactory explanation has yet been provided for the aspects of postnatal behavior that are affected by imprinting and what their evolutionary significance might be [33]. Intriguingly, imprinted genes in plants affect the development and function of the endosperm (which serves as a "placenta" to the embryo) similarly to the effects of maternal and paternal genes as in mammals [34].

As in plants, imprinting in mammals involves DNA methylation [10, 11, 31, 35]. Most imprinted genes have differentially methylated regions (DMRs); some of these differences are acquired in the parental germlines and persist into fetal and adult stages. In early primordial germ cells, imprinted methylation patterns are removed by an unknown mechanism termed erasure [36, 37]. New imprints are introduced by de-novo methylation at later postimplantation stages in male germ cells and after birth during growth of oocytes (Figure 4.1A, B) [10, 11, 31]. It was recently shown that Dnmt3a and b are required in oocytes for maternal methylation imprints to be established, as is Dnmt3L, a protein with homology to 3a and b but without the ability to methylate [38, 39]. This is particularly exciting, because it opens up the possibility of identifying factors that target methylation to specific sequences.

After fertilization, differential methylation of imprinted genes needs to be maintained. This may seem trivial, were it not for the fact that drastic demethylation of the whole genome occurs in mouse preimplantation embryos (Figure 4.1 B). The paternal genome is actively (without a requirement for DNA replication) demethylated by an unknown mechanism only hours after fertilization [6–9]. Despite this genome-wide loss, certain sequences, such as the DMR of the imprinted gene *H19* and perhaps others, seem protected from this demethylation event. It has been argued that the relative scarcity of paternal germline methylation imprints, as compared to maternal ones, is perhaps an outcome of paternal demethylation [40]. From the zygote to the blastocyst stage, replication-dependent passive demethylation also occurs. This is because the preimplantation form of Dnmt1 does not enter the nucleus until the 8-cell stage, at which time it stays in the nucleus for one replication cycle and is then excluded again [41]. This unusual behavior has been investigated recently by knockout of the oocyte- and preimplantation-form of Dnmt1. Lack of Dnmt1o in oocytes resulted in lethality to most but not all offspring at different times of development. Notably, methylation in DMRs of imprinted genes was affected, with the methylated allele losing 50 % of its methylation. This argues that Dnmt1o is required specifically at the 8-cell stage for maintenance of imprinted methylation, but raises the question of what maintains it before and after this specific stage [41].

The unmethylated allele of an imprinted gene has to resist genome-wide de-novo methylation by Dnmt3 a and b after implantation, and it has been suggested that this resistance may be due to a specialized chromatin structure in the unmethylated allele [42]. Some imprinted genes do not show apparent methylation differences, and their imprinted expression appears to be unaffected by a lack of methylation (in *Dnmt1*-embryos), raising the possibility of other imprinting mechanisms, perhaps based on heritable chromatin structures. Recent work has shown that paternal alleles can be heritably modified by egg cytoplasm, apparently without the involvement of methylation [43].

Not all DMRs in germ cells remain so during the reprogramming period in early embryos. For example, *Igf2* DMR2 is methylated in sperm and not in oocytes; after fertilization the paternal methylation is lost [6], but reappears at some stage after implantation. Methylation of this type of DMR might depend on other true germline DMRs within the same cluster of imprinted genes [44]. Thus, the epigenotype within a cluster is regionally controlled; the mechanisms for this are unknown.

Established imprints have to be read in somatic cells that show monoallelic gene expression. Several of these reading mechanisms are beginning to be understood. Methylation of the promoter region on one allele is usually associated with transcriptional silencing of that allele, perhaps due to the action of MBDs that are associated with histone deacetylases (HDACs), the local action of which leads to a condensed chromatin structure that is difficult to transcribe [45]. However, many imprinted genes have DMRs outside their promoters that are actually methylated on the active allele. Some of these regions appear to contain promoters for antisense RNAs. The demethylated DMR transcribes the antisense promoter, and the antisense transcript in turn leads to downregulation of the sense RNA in cis, by an unknown mechanism [46]. Other DMRs of this type contain silencer sequences that may be regulated epigenetically by methylation [47], activator sequences that are methylation sensitive [48], or chromatin boundaries regulated by methylation [49, 50]. Some of these new and interesting epigenetic mechanisms are likely to exist in nonimprinted genes as well.

A

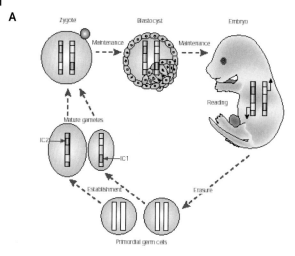

B

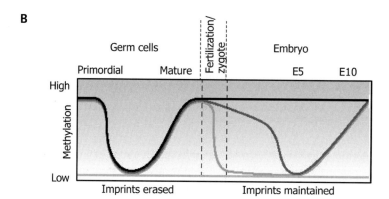

C

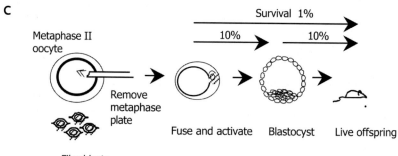

D

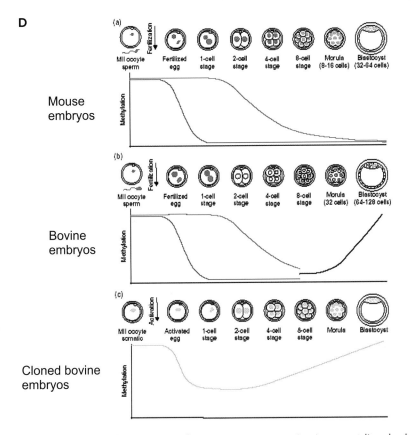

Mouse embryos

Bovine embryos

Cloned bovine embryos

Fig. 4.1 Properties of imprinting and epigenetic reprogramming in mammalian development. **A**. The life cycle of imprints. Methylation imprints are introduced in oocytes and sperm in different imprinted genes (IC1 and IC2 are so-called imprinting centers). Differential methylation is maintained after fertilization and is read and converted into gene-expression differences in embryos and adults. Methylation imprints are erased during development of the primordial germ cells, thus closing the life cycle. **B**. DNA methylation reprogramming. Methylation patterns are reprogrammed genome-wide in primordial germ cells (and imprints are erased at this time) and reestablished in mature gametes. After fertilization, the paternal genome is actively demethylated (blue), whereas the maternal one is passively demethylated during cleavage divisions of the preimplantation embryo (red). De-novo methylation begins in inner cells of the blastocyst. Importantly, imprints are maintained during this period. **C**. Animal cloning. In cloning, differentiated (and highly methylated) nuclei are introduced into enucleated oocytes. This can give rise to live offspring in several mammalian species, but the process is very inefficient. **D**. Deficient reprogramming in cloned embryos. Although some rapid demethylation of donor nuclei occurs initially, further demethylation does not occur, and nuclei end up having abnormally high levels of methylation and a nuclear organization reminiscent of the donor nucleus. This deficient epigenetic reprogramming may be a major cause for the inefficiency of cloning.

4.5
Reprogramming and Cloning

Imprinting methylation differences are erased by demethylation during the development of germ cells and are thus reprogrammed [36, 37]. What is remarkable is that, during preimplantation development, genome-wide demethylation also occurs, despite the need for imprinted methylation differences to persist. The biological purpose of this wave of demethylation, which occurs both by active and passive mechanisms, is not clear. One extreme suggestion is that the paternal demethylation is an "anti-imprinting weapon", which is wielded by the egg against sperm-derived imprints, consistent with the conflict theory [40], but this does not explain the purpose of the passive demethylation. Another view is that demethylation may prime the genome for the widespread transcriptional activation that occurs early in mammals, and there is some support for this view from recent experiments in non-mammalian species such as *Xenopus* [18]. Finally, the highly specialized gametic genomes may need to be reprogrammed, so that the zygotic genome can be totipotent for development of all cell lineages in the embryo [44, 51].

The significance of reprogramming in the preimplantation embryo could be tested in cloned embryos (Figure 4.1 C, D). Cloning of mammals from somatic cells is now possible in several species [44, 51]. The somatic, presumably differentiated, nucleus is transferred into an enucleated oocyte that is subsequently activated to undergo embryonic development. Although viable offspring can be obtained in various mammalian species, the success rate is exceedingly low. Typically only 1% of reconstituted zygotes develop to term (Figure 4.1 C). Heavy losses occur during preimplantation development, and more losses occur after implantation and are often accompanied by placental abnormalities, most typically placental hyperplasia. It is unclear whether these placental abnormalities are a cause of developmental failure.

Several explanations have been suggested for the problems with cloning, including difficulties in epigenetic reprogramming. Presumably, the somatic donor nuclei are highly methylated and possess all the chromatin modifications characteristic of differentiated cells. However, it is not known how stable these epigenetic modifications are when cells are derived in culture. In the mouse, embryonic stem (ES) cells are being used successfully in cloning experiments [51]. This poses a problem, because methylation patterns in imprinted genes are unstable in ES cells [52]; this instability is retained in cloned embryos and may well lead to developmental problems in these embryos.

Several studies have recently begun to address the question of if and how DNA methylation patterns are reprogrammed in cloned embryos [53–56]. These studies found that, although some reprogramming occurs (with perhaps loss of DNA methylation as early as the reconstituted zygote), in most cloned morulae and blastocysts the levels of DNA methylation were far too high (more characteristic of somatic nuclei than of embryonic ones), and the nuclear organization of methylation patterns was reminiscent of the somatic donor nuclei (Figure 4.1 D). This was particularly obvious in the trophectoderm cells of the blastocysts, which are normally quite undermethylated in several mammalian species. Thus, the aberrantly high levels of

DNA methylation in the trophectoderm lineage, which goes on to form important parts of the placenta, could be responsible for the consistently observed placental abnormalities in cloned embryos. Intriguingly, a very recent study found widespread downregulation of gene expression in placentas but not in embryos of cloned mice [57]. In addition to DNA methylation, of course, other chromatin-reprogramming processes need to be examined in cloned embryos.

4.6
Epimutations and Epigenetic Inheritance

Epimutations are abnormal changes in epigenetic modifications, just as mutations are changes in the DNA sequence. Epimutations may arise from mutations in trans-acting factors, such as methylases that are important in epigenetic patterns. Epimutations may also arise spontaneously at a certain frequency; this frequency may depend on other genetic and environmental factors. For example, a few patients with Beckwith Wiedemann syndrome or Prader Willi-Angelman syndromes have epimutations in imprinted genes [58–60]. These may arise from the occasional failure to erase or to properly establish the imprint in germ cells, or alternatively, from defects in embryos. Early embryos and their stem cells (ES cells) seem particularly prone to acquiring epimutations when exposed to manipulations or adverse culture conditions [51, 52], and these are not normally reversible during somatic development. Of course, epimutations are not limited to imprinted genes.

Imprinted methylation patterns are efficiently reprogrammed in the germ cells as are, apparently, methylation patterns in genes of some other cell types [36, 37]. But is all epigenetic information erased in germ cells, so that none is inherited by future generations? This seems unlikely. First, a number of observations, particularly from human disease pedigrees, suggest transgenerational effects that may be based on this type of epigenetic inheritance. Further, a variety of transgenes in the mouse have methylation patterns that are not fully reversed on passage through the germ line, resulting in epigenetic inheritance of methylation patterns. Aberrant inherited methylation patterns and growth phenotypes have also been described in certain experiments on mouse nuclear transplantation [61]. Recently, epigenetic inheritance, through the female germline, of methylation and expression patterns of the Agouti viable yellow gene allele has been observed [62]. Methylation occurs upstream of the Agouti gene, where an intracisternal A particle (IAP) transposable element is inserted and leads to ectopic transcription of A and abnormal phenotypes when unmethylated, but to a normal phenotype when methylated. Mothers that are highly methylated in this IAP have offspring that are on average more methylated, whereas mothers that are demethylated have offspring that are relatively demethylated. Indeed, a recent study found that DNA methylation patterns in IAP elements are not completely reprogrammed in the germ line [37]; thus, epigenetic inheritance may result from failure to reprogram in the germ line, in the early embryo, or in both.

4.7

Epigenomics: The Future

Many of the examples in this chapter and this book highlight the ever-increasing significance and impact of epigenetics in biology and medicine. What needs to happen next is a systematic cataloguing of different epigenetic modifications genome wide and also in various developmental and tissue situations (see Chapters 4 and 5 in this volume). This may seem like a daunting task, but methods for high-throughput analysis are already emerging. For DNA methylation; the field is led by Epigenomics AG, who prepare oligoDNA arrays designed against genomic DNA sequences treated with the chemical bisulfite, which deaminates cytosines (to be converted to thymidines) when they are unmethylated but not when methylated. Following multiplex PCR and fluorescent labeling, signals that are detected on the arrays indicate whether a particular sequence stretch is methylated or not. Initial applications of this technique to cancer biology look very promising; certain tumor classes can be identified, and some methylation sites were predictive of tumor progression [63]. Other systems for high-throughput analysis of DNA methylation have been described [64], and systems based perhaps on chromatin immunoprecipitation (ChIP) should allow, at least in principle, the high-throughput analysis of chromatin modifications and binding factors.

If disease states are critically dependent on epigenetic defects (some have already turned out to be), thought should be given to epigenetic intervention in therapy as well. This would be particularly relevant for situations in which a small number of specific genes are deregulated. For example, reactivation of methylated and silenced tumor suppressors or mutator genes can significantly alter tumor phenotype. Thus, targeted epigenetic reactivation or inactivation in tumors could be a highly promising therapeutic approach [65].

In a similar vein, the application of cloning and stem cell therapy has now been shown to be feasible in principle [66, 67]. Eventually however, through increased understanding of the biology of reprogramming, it may be possible to manipulate reprogramming processes in differentiated cells, thus returning them to an undifferentiated stem cell state for therapy, without a need for cloning.

4.7

Conclusions

With the completion of the genome sequence in several mammals, mammalian epigenetics and epigenomics will take center stage in the functional analysis of genome function, its interaction with the environment, and its impact on human disease. We can speculate that environmental influences on epigenetic functions will turn out to be an important component of phenotypic variation in natural populations, possibly with transgenerational effects and, therefore, short-term adaptive potential. An equally important role can be envisaged for epigenetic modifications in aging and many disease syndromes. The future potential of "epigenetic engineering", that is, the experimental or therapeutic alteration of epigenetic information in the cell, holds great excitement for scientists and physicians alike.

References

1 Epigenetics issue. *Science* **2001**, *293*, 1063–1105.

2 Lyko, F.; Ramsahoye, B. H.; Jaenisch, R., *Nature* **2000**, *408*, 538–540.

3 Bestor, T. H., *Hum. Mol. Genet.* **2000**, *9*, 2395–2402.

4 Bird, A., *Genes. Dev.* **2002**, *16*, 6–21.

5 Okano, M.; Bell, D. W.; Haber, D. A.; Li, E., *Cell* **1999**, *99*, 246–257.

6 Oswald, J.; Engemann, S.; Lane, N.; Mayer, W.; Olek, A.; Fendele, R.; Dean, W.; Reik, W.; Walter, J., *Curr. Biol.* **2000**, *10*, 475–478.

7 Mayer, W.; Niveleau, A.; Walter, J.; Fundele, R.; Haaf, T., *Nature* **2002**, *403*, 501–502.

8 Santos, F.; Hendrich, B.; Reik, W.; Dean, W., *Dev. Biol.* **2002**, 172–182.

9 Cedar, H.; Verdine, G. L., *Nature* **1999**, *397*, 568–569.

10 Reik, W.; Walter, J., *Nature Rev. Genet.* **2001**, *2*, 21–32.

11 Sleutels, F.; Barlow, D. P., *Adv. Genet.* **2002**, *46*, 119–163.

12 Heard, E.; Avner, P., *Nature Rev. Genet.* **2001**, *2*, 59–67.

13 Chen, R. Z.; Pettersson, U.; Beard, C.; Jackson-Grusby, L.; Jaenisch, R., *Nature* **1998**, *395*, 89–93.

14 Lengauer, C.; Kenzler, K. W.; Vogel-stein, B., *Proc. Nat. Acad. Sci. USA* **1997**, *94*, 2545–2550.

15 Riggs, A. D., *Cytogenet. Cell Genet.* **1975**, *14*, 9–25.

16 Holliday, R.; Pugh, J. E., *Science* **1975**, *187*, 226–232.

17 Jackson-Grusby, L.; Beard, C.; Posse-mato, R.; Tudor, M.; Fambrough, D.; Csankoyszki, G.; Dausman, J.; Lee, P.; Willson, C.; Lander, E.; Jaenisch, R., *Nature Genet.* **2001**, *27*, 31–39.

18 Walsh, C. P.; Bestor, T. H., *Genes Dev.* **1999**, *13*, 26–34.

19 Stancheva, I.; Meehan, R. R., *Genes Dev.* **2000**, *14*, 313–327.

20 Walsh, C. P.; Chaillet, J. R.; Bestor, T. H., *Nature Genet.* **1998**, *20*, 116–117.

21 Yoder, J. A.; Walsh, C. P.; Bestor, T. H., *Trends Genet.* **1997**, *13*, 335–340.

22 Barlow, D. P., *Science* **1993**, *260*, 309–310.

23 Jones, P. A.; Takai, D., *Science* **2001**, *293*, 1068–1070.

24 Roundtree, M. R.; Bachman, K. E.; Herman, J. G.; Baylin, S. B., *Oncogene* **2001**, *20*, 3156–3165.

25 Bird, A. P.; Wolffe, A. P., *Cell* **1999**, *99*, 451–454.

26 Turner, B. M., *BioEssays* **2000**, *22*, 836–845.

27 Jenuwein, T.; Allis, C. D., *Science* **2001**, *293*, 1074–1080.

28 Richards, E. J.; Elgin, S. C., *Cell* **2002**, *108*, 489–500.

29 Tamaru, T.; Selker, E. U., *Nature* **2001**, *414*, 277–283.

30 Jackson, J. P.; Lindroth, A. M.; Cao, C.; Jacobson, S. E., *Nature* **2002**, *416*, 556–560.

31 Ferguson-Smith, A. C.; Surani, M. A., *Science* **2001**, *293*, 1086–1089.

32 Moore, T.; Haig, D., *Trends Genet.* **1991**, *7*, 47–49.

33 Isles, A.; Wilkinson, L., *Trends Cogn. Sci.* **2000**, *4*, 309–318.

34 Baroux, C.; Spillane, C.; Grossnik-laus, U., *Adv. Genet.* **2002**, *46*, 165–214.

35 Adams, S.; Vinkenoog, R.; Spiel-man, M.; Dickinson, H. G.; Scott, R. J., *Development* **2000**, *127*, 2493–2450.

36 Lee, J.; Inoue, K.; Ono, R.; Ogonu-ki, N.; Kohda, T.; Kaneko-Ishino, T.; Ogura, A.; Ishino F., *Development* **2002**, *129*, 1807–1817.

37 Hajkova, P.; Erhardt, S.; Lantz, N.; Haaf, T.; El-Maari, O.; Reik, W.; Walter, J.; Surani, M. A., *Mech. Dev.* **2002**, *117*, 15–22.

38 Bourchis, D.; Xu, G. L.; Lin, C. S.; Bollman, B.; Bestor, T. H., *Science* **2001**, *294*, 2536–2539.

39 HATA, K.; OKANO, M.; LEI, H.; LI, E., *Development* **2002**, *129*, 1983–1993.

40 REIK, W.; WALTER, J., *Nature Genet.* **2001**, *27*, 255–256.

41 HOWELL, C. Y.; BESTOR, T. H.; DING, F.; LATHAM, K. E.; MERTINEIT, C.; TRASLER, J. M.; CHAILLET, J. R., *Cell* **2001**, *104*, 829–838.

42 FEIL, R.; KHOSLA, S. *Trends Genet.* **1999**, *15*, 431–435.

43 PICKARD, B.; DEAN, W.; ENGEMANN, S.; BERGMANN, K.; FUERMANN, M.; JUNG, M.; REIS, A.; ALLEN, N.; REIK, W.; WALTER, J., *Mech. Dev.* **2001**, *103*, 35–47.

44 REIK, W.; DEAN, W.; WALTER, J., *Science* **2001**, *293*, 1089–1093.

45 DREWELL, R. A.; GODDARD, C. J.; THOMAS, J. O.; SURANI, M. A., *Nuc. Acids Res.* **2002**, *30*, 1139–1144.

46 SLEUTELS, F.; ZWART, R.; BARLOW, D. P., *Nature* **2002**, *415*, 810–813.

47 EDEN, S.; CONSTANCIA, M.; HASHIMSHONY, T.; DEAN, W.; GOLDSTEIN, B.; JOHNSON, A. C.; KESHET, I.; REIK, W.; CEDAR, H., *EMBO J.* **2001**, *20*, 3518–3525.

48 MURRELL, A.; HEESON, S.; BOWDEN, L.; CONSTANCIA, M.; DEAN, W.; KELSEY, G.; REIK, W., *EMBO R.* **2001**, *2*, 1101–1106.

49 HARK, A. T.; SCHOENHERR, C. J.; KATZ, D. J.; INGRAM, R. S.; LEVORSE, J. M.; TILGHMAN, S. M., *Nature* **2000**, *405*, 486–489.

50 BELL, A. C.; FELSENFELD, G., *Nature* **2000**, *405*, 482–485.

51 RIDEOUT, W. M. III; EGGAN, K.; JAENISCH, R., *Science* **2001**, *293*, 1093–1098.

52 DEAN, W.; BOWDEN, L.; AITCHISON, A.; KLOSE, J.; MOORE, T.; MENESES, J.; REIK, W.; FEIL, R., *Development* **1998**, *125*, 2273–2282.

53 DEAN, W.; SANTOS, F.; STOJKOVIC, M.; ZAKHARTCHENKO, V.; WALTER, J.; WOLF, E.; REIK, W., *Proc. Nat. Acad. Sci. USA* **2001**, *98*, 13734–13738.

54 BOURCHIS, D.; LE BOURHIS, D.; PATIN, D.; NIVELEAU, A.; COMIZZOLI, P.; RENARD, J. P.; VIEGAS-PEQUIGNOT, E., *Curr. Biol.* **2001**, *11*, 1542–1546.

55 KANG, Y. K.; KOO, D. B.; PARK, J. S.; CHOI, Y. H.; CHUNG, A. S.; LEE, K. K.; HAN, Y. M., *Nature Genet.* **2001**, *28*, 173–177.

56 KANG, Y. K.; PARK, J. S.; KOO, D. B.; CHOI, Y. H.; KIM, S. U.; LEE, K. K.; HAN, Y. M., *EMBO J.* **2002**, *21*, 1092–1100.

57 INOUE, K.; KOHDA, T.; LEE, J.; OGONUKI, N.; MOCHIDA, K.; NOGUCHI, Y.; TANEMURA, K.; KANEKO-ISHINO, T.; ISHINO, G.; OGURA, A., *Science* **2002**, *295*, 297.

58 REIK, W.; BROWN, K. W.; SCHNEID, H.; LE BOUC, Y.; BICKMORE, W.; MAHER, E. R., *Hum. Mol. Genet.* **1995**, *4*, 2379–2385.

59 BUITING, K. ET AL., *Amer. J. Hum. Genet.* **1998**, *63*, 170–180.

60 SMILINICH, N. D., et. al., *Proc. Nat. Acad. Sci. USA* **1999**, *96*, 8064–8069.

61 ROEMER, I.; REIK, R.; DEAN, W.; KLOSE, J., *Curr. Biol.* **1997**, *7*, 277–280.

62 MORGAN, H. D.; SUTHERLAND, H. G.; MARTIN, D. I.; WHITELAW, E., *Nature Genet.* **1999**, *23*, 314–318.

63 ADORJAN, P.; DISTLER, J.; LIPSCHER, E.; MODEL, F.; MULLER, J.; PELET, C.; BRAUN, A.; FLORI, A. R.; GUTIG, D.; GRABS, G.; HOWE, A.; KURSAR, M.; LEUSCHE, R.; LEU, E.; LEWIN, A.; MAIER, S.; MULLER, V.; OTTO, T.; SCHOLZ, C.; SCHULZ, W. A.; SEIFERT, H. H.; SCHWOPE, I.; ZIEBARTH, H.; BERLIN, K.; PIEPENBROCK, C.; OLEK, A., *Nucleic Acids Res.* **2002**, *30*, 21.

64 TOMPA, R.; McCULLUM, C. M.; DELROW, J.; HENIKOFF, J. G.; VAN STEENSEL, B.; HENIKOFF, S., *Curr. Biol.* **2002**, *12*, 65–68.

65 REIK, A.; GREGORY, P. D.; URNOV, F. D., *Curr. Opin. Genet. Dev.* **2002**, *12*, 233–242.

66 RIDEOUT, W. M.; HOCHEDLINGER, K.; KYBA, M.; DALEY, G. Q.; JAENISCH, R., *Cell* **2002**, *109*, 17–27.

67 BYRNE, J. A.; SIMONSSON, S.; GURDON, J. B., *Proc. Natl. Acad. Sci. USA* **2002**, *99*, 6059–6063.

5
At the Controls:
Genomic Imprinting and the Epigenetic
Regulation of Gene Expression

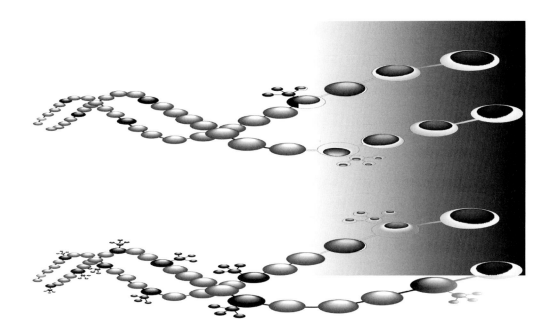

5

At the Controls: Genomic Imprinting and the Epigenetic Regulation of Gene Expression

ANNE FERGUSON-SMITH

Summary

Epigenetic modifications are molecular "flags" on chromosomes that play important roles in regulating genome function, including activation and repression of some genes. Epigenetic modifications that influence gene expression include DNA methylation and modifications to core histones that affect chromatin structure and function over both short and long distances. Genomic imprinting is an epigenetic marking process that causes some genes to be expressed according to their parental origin. Over the past decade, understanding of how imprinted genes are regulated has provided an excellent model for determining the roles of epigenetic modifications in controlling gene activity and repression. This is because within the same cell, one allele of an imprinted gene is expressed, while at the same time the other allele is silent. Many imprinted genes are located in clusters, and studies on the mechanisms of imprinting have identified elements that are differentially modified on the two parental chromosomes and can act locally on the adjacent gene or over a longer range, affecting the activity of multiple imprinted genes in the cluster. Comparing modifications on the two parental alleles provides major insight into how epigenetic control is initiated and maintained in a heritable fashion over multiple cell divisions. This information can then be applied to understanding the interactions between the modified genome and the factors that directly influence transcription. Imprinted genes can exhibit DNA methylation on either the active or the repressed allele but not both, and sometimes, although not always, DNA methylation can be correlated with modifications to the associated histones. In some instances these modifications not only affect the ability of regulatory proteins to interact with the chromatin, but also regulate the transcription of noncoding antisense RNAs that are required for silencing gene-coding transcripts of the sense strand. Challenges now facing researchers include identifying more key elements required for epigenetic control and determining whether general genomic features that control imprinting are common to multiple epigenetically regulated domains. In addition, understanding how the noncoding RNAs function and determining the relationships between DNA methylation and modifications to core histones remain to be elucidated. Furthermore, there is considerable interest in understanding how these mechanisms evolved. In addition to genomic imprinting, similar epigenetic mechanisms are involved in inactivation of the X chromosome in female mammals and in silencing repetitive "parasitic" elements that make up a significant portion of the mammalian genome. The application of epigenetic control mechanisms to the regulation of nonimprinted autosomal genes is a controversial area that remains to be comprehensively addressed.

5.1
Introduction

Genome function is modulated by epigenetic factors that can affect several features, including chromosome architecture and function, chromatin structure, and gene expression. Epigenetic modifications can be considered to be molecular "flags" that program the genome to form a particular chromatin conformation, interact with regulatory and structural proteins, and allow chromosomes to function appropriately during all phases of the cell cycle. During the lifetime of an organism, these modifications provide a dynamic, heritable, and reversible method to affect genome function without changing the DNA sequence. The epigenetic factors involved are usually modifications to the DNA or chromatin, and the best studied include DNA methylation [1] and the modification of core histones [2]. Epigenetic modifications provide functional flexibility to the genomic template on which they act.

DNA methylation is the covalent modification of DNA at the cyclic carbon-5 of a cytosine residue; 5-methyl cytosine is found in many eukaryotic organisms, albeit to different extents. Extensive background information on DNA methylation, including the properties of the methylation machinery and the developmental stages when heritable methylation marks undergo reprogramming, can be found elsewhere in this book (see Chapters 1 and 4). In mammals, methylated cytosine is found predominantly at CpG dinucleotides, and in differentiated cells approximately 75% of CpGs are methylated. Most CpGs are located in CG-rich stretches of DNA known as CpG islands, which are usually, although not always, localized in the 5′ regions of genes. In general, CpG islands remain unmethylated during development and in the adult. However, extensive studies of CpG-island methylation have demonstrated their broad involvement in a variety of normal and disease processes, including genomic imprinting, X-inactivation, cancer, and aging (see Chapter 6). Furthermore, evidence from mammalian and nonmammalian vertebrate species is emerging, that DNA methylation plays an important role in regulating tissue- and stage-specific genes during development. For example, in *Xenopus* embryos DNA methylation regulates the timing of the mid-blastula transition [3] – a critical time in which there is a switch from a requirement for maternally encoded transcripts to the transcription of embryonic genes, which are required for further development. In most instances, DNA methylation is associated with transcriptional silencing. The reversible and stable nature of DNA methylation makes it a reasonable mechanism for regulating genes in a tissue-specific or developmentally specific manner; nonetheless, the role of DNA methylation in the control of developmentally regulated mammalian genes (those that are not subject to genomic imprinting) has been controversial. Although it has long been hypothesized that CpG methylation might control tissue-restricted gene expression during mammalian development, a definitive correlation between methylation and the silencing of developmentally regulated genes is lacking. Both direct and indirect evidence suggest that, although many developmentally regulated genes are not regulated by methylation [4], others clearly are [5, 6]. Furthermore, mouse fibroblast cells lacking the methylating enzyme DNA methyltransferase I (Dnmt1) show an overall 10% increase in gene activity, including that of some tissue-specific genes [7]. Although indirect mechanisms cannot be ruled out here, the findings from this study are consistent with data emerging for individual genes assayed for developmental changes due to methylation.

5.2
Genomic Imprinting

It is well established that DNA methylation contributes to the regulation of expression of a subset of remarkable developmentally regulated genes, the imprinted genes. Genomic imprinting is an epigenetic marking process that causes genes from only one of the two parental chromosome homologues to be expressed heritably (Fig. 5.1). Hence, the two parental genomes exhibit functional asymmetry due to the presence of imprinted genes that are expressed mono-allelically [8]. For example, normally all mammals, the insulin-like growth factor II gene is expressed from the paternally inherited chromosome while the maternally inherited allele remains silent.

Imprinting is a normal process that plays an important role in regulating genes involved in prenatal growth, postnatal behavior, and the development of particular cell lineages. The presence of imprinted genes is the reason why mammals do not support parthenogenetic (bimaternal) or androgenetic (bipaternal) development. So, how does the transcriptional machinery of the cell tell the difference between the two parental alleles of an imprinted gene, resulting in the expression of one and silencing of the other? What is the imprinting mechanism? Over the past ten years, it has become clear that this differential expression is achieved by specific epigenetic

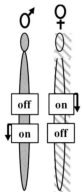

Fig. 5.1   Schematic representation of a pair of closely linked imprinted genes on the paternally and maternally inherited chromosome homologs. Genomic imprinting causes genes to be expressed according to parental origin. This results in one allele of an imprinted gene being expressed (on) and the other allele being silent (off). This allele-specific control of transcription is regulated by epigenetic modifications to the DNA and chromatin.

modifications that are different on the two alleles of an imprinted gene and depend on their parental origin. Thus, because a normal cell contains both an active and a repressed copy of the same gene, we are provided with an excellent model system for studying the mechanisms by which epigenetic modifications control gene expression. Knowledge generated from expression studies of imprinted genes continues to be applied to other epigenetically regulated processes. Indeed, much of our understanding of the mechanisms of epigenetic transcriptional control comes from the study of imprinted genes and, with the exception of the parental origin-specific behavior of imprinted genes, generalized mechanisms are emerging suggesting that we can learn about more wide-spread epigenetic controlling mechanisms through the study of imprinted genes. This chapter reviews how epigenetic modifications regulate genomic imprinting.

5.2.1
The Role of DNA Methylation in Imprinted Gene Expression

Initial parental-origin-specific imprinting signals are likely to be established during gametogenesis. In the germline, in primordial germ cells, old imprints are erased and new parental-specific imprints are established. This is the first phase of so-called nuclear reprogramming, obviously occurring in a place where the two parental genomes are isolated from each other. Precise modifications and mechanisms associated with the establishment of new imprints in the gametes are not well understood; however, they differ in their target sequences in oogenesis and spermatogenesis and may even involve different mechanisms. After fertilization and before implantation, a second wave of nuclear reprogramming occurs, during which the two parental genomes are once again subject to changes in their epigenetic status. The epigenetic changes associated with nuclear reprogramming involve changes in DNA methylation and in the modifications of core histones. The dynamic epigenetic changes associated with nuclear reprogramming are discussed in Sect. 4.

DNA methylation is involved in the mechanism of genomic imprinting. Imprinted genes usually have CpG islands in their promoters and often elsewhere. Importantly, intergenic and noncoding regions in the vicinity of imprinted genes also contain CpG islands. CpG islands that are hypermethylated on one of the two parental alleles and unmethylated on the other are known as differentially methylated regions (DMRs). This differential methylation can serve as an epigenetic mark that allows the two chromosome homologues to be distinguished from each other. Imprinted domains contain two types of DMR. The first type is established in the germline and results in a specific methylation difference that is inherited from the egg or sperm. These are called germline or gametic imprints. The second type occurs after fertilization. Although it is generally assumed that these DMRs are involved in the local control of allele-specific gene expression, little is known about their primary importance in imprinting. We do not know whether these somatic imprints are achieved as a direct consequence of cis-acting germline imprints or are an indirect secondary effect of gene activation or repression after fertilization. Germline imprints have been termed imprinting controlling regions (ICRs), imprinting controlling elements (ICEs), or imprinting centers (ICs), if they regulate the monoallelic activity of several imprinted genes along the same chromosome (i. e., cis).

5.2.3
Organization of Imprinted Genes

To date, approximately 50 imprinted genes have been identified, which map to around 12 locations within the genome [9]. Many, although not all, are located in clusters that contain genes expressed from maternally inherited alleles alongside genes that are expressed from paternally inherited alleles. This has implications for their control, since imprints act over short distances, affecting gene-specific regulatory elements such as promoters, and also over longer distances at sites that affect the activity or repression of several imprinted genes within the cluster. Although imprinted domains can contain closely linked reciprocally imprinted genes, the methylation at each domain is generally associated with only one of the two parental chromosomes and hence is associated with both active and repressed alleles (see below). Most of our understanding of how imprinted genes are regulated comes from the study of three imprinted domains. The first is the domain on mouse distal chromosome 7/human chromosome 11p that contains a cluster of at least 12 developmentally regulated imprinted genes within a region of approximately 800 kb [10]. The function of these imprinted genes has been studied to some extent and, to date, it appears that they act in the control of prenatal growth, with many expressed in the placenta [11–16]. In humans, defects in the imprinting of these genes is associated with the somatic overgrowth disorder Beckwith-Wiedemann Syndrome (BWS), and BWS patients also have an increased incidence of childhood tumors [17]. Another imprinted domain encompassing 3–4 Mb is located on central mouse chromosome 7/human chromosome 15q. In contrast to the genes in the BWS cluster, these genes play a role in neural function and are predominantly expressed and imprinted in the brain [18]. Defects in the imprinting of these genes are associated with two distinct neurological disorders, the Prader-Willi (PWS) and Angelman (AS) syndromes [19]. The human diseases associated with imprinted loci are discussed in Sect. 6. A third imprinted locus on mouse chromosome 17 contains the imprinted *Igf2r* gene and a small cluster of genes that show tissue-specific imprinting in the placenta [20]. Again, these imprinted genes appear to function in prenatal growth control. It appears that these three loci are regulated differently from each other and, to date, except for a role of DNA methylation in distinguishing the two parental alleles, no common themes in the mechanisms of their regulation are obvious. This is relevant to those in the field interested in the evolution of the imprinting mechanism and raises the question of whether any common genomic features confer epigenetic control on imprinted domains. Comparative genomic and mechanistic analyses of these and additional imprinted domains should provide answers to this question. In general, gene imprinting appears to be conserved between different mammalian species. One notable exception is the *Igf2r* locus, which in humans does not show consistent imprinting [21]. Not all imprinted genes are found in clusters [9]. These individual genes in imprinted "microdomains" are currently in the minority and are not discussed here. Nonetheless, comparative analysis of the mechanisms regulating imprinted microdomains versus imprinted domains in gene clusters has the potential to provide insight into both short- and long-range mechanisms of epigenetic control.

A role for DNA methylation in genomic imprinting was established by studying embryos lacking the enzyme required for maintaining DNA methylation, DNA methyltransferase 1 (*Dnmt1*). In these animals monoallelic expression was lost, and several silenced alleles became activated, resulting in biallelic expression. Thus, methylation is required for the silencing of some imprinted genes. However, in the *Dnmt1* mutant, other imprinted genes also became silenced. This suggested that DNA methylation is required for the activity of some imprinted genes [22]. Therefore, as far as imprints are concerned, DNA methylation can be associated with both the silencing and activity of imprinted alleles. To date, only one imprinted gene (*Mash2*) has been identified whose monoallelic activity is retained in the absence of *Dnmt1* [23].

5.2.4
The Mechanism of Imprinting at the Mouse *Igf2r* Imprinted Domain Requires a Cis-acting Noncoding Antisense Transcript Regulated by DNA Methylation (Fig. 5.2 a)

The mouse *Igf2r* gene is located in a 400-kb imprinted domain and encodes a mannose-6-phosphate receptor that also functions to bind IGFII and target it to lysosomes for degradation [24]. It is expressed from the maternally inherited allele and silent on the paternally derived copy. Its promoter is a CpG island that becomes methylated late in gestation on the already silent paternal allele. Therefore in this case, the paternal allele can be silenced in the absence of promoter methylation. A second DMR acts as an ICR and is located within a CpG island in the second intron of *Igf2r*. In contrast to the promoter DMR, this CpG island carries a germline imprint that is inherited from the egg in a hypermethylated state [25]. On the unmethylated paternally inherited allele, this DMR is associated with an imprinted antisense transcript whose expression is required for transcriptional silencing of *Igf2r* from the paternal allele. On the methylated maternal allele, the intronic CpG island is methylated and antisense transcription does not occur. Interestingly, on the paternal allele, this antisense transcript is also required for silencing two other imprinted genes located upstream of the antisense transcript. These two genes, *Slc22a2* and *Slc22a3*, encode the organic cation transporter proteins and are expressed from the maternally inherited allele in the placenta. Their imprinting is specific to the placenta and neither gene is expressed in the embryo. Truncation of the *Igf2r* antisense transcript results in loss of imprinting of *Igf2r*, *Slc22a2*, and *Slc22a3* on the paternally inherited allele and indicates that this transcript is required bidirectionally for silencing three paternally inherited protein-coding genes [26]. This model has parallels with X inactivation in female mammals, in which expression of a noncoding *Xist* RNA inactivates the X chromosome from which it is expressed [27].

5.2.5
Imprinting at the Mouse Distal Chromosome 7 Domain Is Regulated by at Least Two Imprinting Control Regions (ICRs) Acting on Different Sets of Genes (Fig. 5.2 b & c)

Analysis of the mechanisms regulating the monoallelic expression of a cluster containing at least 12 imprinted genes on distal mouse chromosome 7 has been of interest because of the association of this locus with BWS, a human genetic disorder that exhibits parental-origin effects in its pattern of inheritance ([17] and Chapter 6). More detailed analysis in both mouse and human shows that the region can be subdivided into two imprinted subdomains, one containing the well characterized *Igf2-H19* domain, and the other, the very large *Kcnq1* gene (Fig. 5.2 b and c). Imprinting at these two subdomains is regulated by different and independent mechanisms (see below), although functional interactions between their gene products have been reported [28].

The paternally expressed *Igf2* gene encoding the insulin-like growth factor II protein is located 90 kb upstream from the maternally expressed *H19* gene, which is transcribed into a highly abundant noncoding RNA of unknown function. During development the two genes are expressed monoallelically in the same tissues, except in parts of the brain, where *Igf2* expression is biallelic. Except in the brain, the two genes use common regulatory elements for their expression and their imprinting [29]. In *Dnmt1* mutant mouse embryos, the usually silent paternal *H19* allele becomes activated, and the active paternal *Igf2* allele becomes silenced [22]. This result indicates that DNA methylation is a positive regulator of *Igf2* and a negative regulator of *H19* and was one of the first indications that the mechanism of imprinting of the two genes might be mechanistically linked.

In the *Igf2-H19* domain, local DMRs involved in the regulation of imprinting in the domain have been identified. *Igf2-H19* includes five DMRs, of which four are hypermethylated on the paternal allele and hypomethylated on the maternal allele (Fig. 5.2 b). The one DMR that shows maternally inherited methylation is located in a CpG-island promoter upstream from the fetal *Igf2* promoters, which drives imprinted expression of a placenta-specific *Igf2* transcript [13]. The fetal promoters themselves are embedded within a CpG island that is not differentially methylated on the two parental alleles [30]. This complete lack of methylation is in contrast to the CpG-island promoter of the *H19* gene, in which post-fertilization methylation is associated with the silent paternal allele [31].

Much attention has focused on a DMR located upstream of *H19* where paternal allele-specific methylation is inherited from the sperm (reviewed in [29]). This DMR is an ICR and contains several discrete sequence elements, including a repeated binding site for the zinc-finger DNA binding protein CTCF. This DMR is postulated to have two functions: as a germline imprint mark and as an insulator element in somatic cells. On the paternal chromosome, it acts as the germline-imprinting signal, because when this DMR is deleted, *H19* becomes activated and *Igf2* becomes silenced; hence imprinting is lost. Upon deletion of this DMR from the maternal chromosome, *H19* expression is reduced considerably and *Igf2* is activated. The two genes share common enhancers that are located downstream of *H19*.

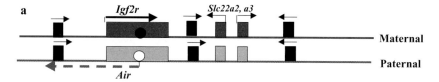

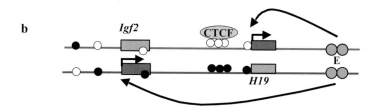

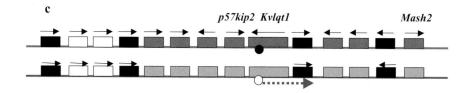

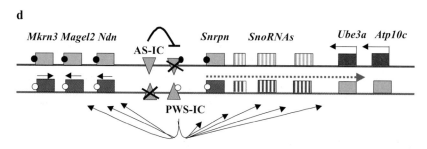

Fig. 5.2 Genomic imprinting is regulated by differential methylation. Circles represent DMRs: black: methylated; and white: unmethylated. Imprinted genes expressed from the maternally inherited allele are shown in red, and those expressed from the paternally inherited chromosome are blue. Genes that are not imprinted are black, and white boxes denote genes of unknown imprinting status. Grey boxes represent the silent alleles of an imprinted gene. Antisense transcripts are represented by dotted lines, and all identified to date are paternally expressed (blue). **a)** Imprinting at the mouse *Igf2r* locus is regulated by differential methylation at an intronic differentially methylated domain (DMR) controlling the expression of an antisense transcript (*Air*). The *Air* transcript is expressed from the unmethylated paternal allele and results in the silencing of *Igf2r* and adjacent downstream imprinted transcripts. *Air* therefore acts bidirectionally to regulate imprinting of multiple genes in the domain. **b)** Imprinting at the *Igf2-H19* domain is regulated by enhancer/insulator/silencer interactions. Five DMRs have been identified at this locus. From left to right, DMR0 is at

The choice of which gene the enhancers activate is regulated by CTCF binding to the DMR 2kb upstream of *H19*. When methylated on the paternal allele, CTCF cannot bind and the enhancers can act on the *Igf2* promoters. On the unmethylated maternal allele, CTCF binds and acts to insulate the enhancers from the *Igf2* gene and to drive expression from the unmethylated *H19* promoter [29].

These studies implicate the differentially methylated insulator upstream of *H19* as the key regulator of imprinting in the domain and suggest that the *Igf2* locus has little influence on regional control. This is actually not true: A DMR located upstream of *Igf2*, between the placental and fetal promoters, does reduce methylation on the silent maternal allele relative to the paternal allele (Fig 5.2 b; DMR1). This unmethylated DMR may bind a repressor element associated with silencing of the maternal allele. So, is the maternal *Igf2* silenced by the downstream insulator or by the upstream repressor? To test this, the maternally inherited DMR1 was deleted. Although paternal inheritance of the deletion has no effect on the normal *Igf2* activity from the fetal promoters, maternal transmission of the deletion results in loss of imprinting of *Igf2*, activating it in several mesodermal tissues, although not to levels as high as the paternal allele. There was no effect on *H19* expression. These results suggest that downstream mesodermal enhancers are not fully insulated by CTCF binding and that this upstream element is also required in mesodermal tissues such as muscle [32, 33]. Additional studies have shown that DMR1 binds the repressor element GCF2 and that methylation abrogates its binding [34]. Although most evidence links DNA methylation to gene silencing, this *Igf2* DMR1 is an example of

---

Fig. 5.2 (continued)

the promoter regulating the placenta-specific *Igf2* transcript. DMR1, when unmethylated, is required for silencing the maternal *Igf2* allele in mesodermal tissues via the binding of a repressor element. DMR2 is located in the last exon of *Igf2* and contributes to normal expression levels on the paternal allele via an unknown mechanism. The intergenic DMR upstream of *H19* is an imprinting-control region (ICR) and is required for germline imprinting of both *Igf2* and *H19*. It also binds the insulator protein CTCF on the unmethylated maternal allele and this insulates the downstream enhancers (E-circles) from the *Igf2* promoters, allowing them to drive *H19* expression from the maternal allele. When methylated on the paternal allele, CTCF cannot bind to the ICR and the enhancers can drive paternal *Igf2* activity. The *H19* promoter is also a DMR, becoming methylated on the paternal allele after fertilization. **c)** Imprinting at the *Kvlqt1* domain is controlled, at least in part, by a CpG island that is methylated on the maternally inherited chromosome and unmethylated on the paternal allele. When unmethylated, an antisense transcript is expressed and is required for silencing the paternal alleles of *Kvlqt1* and *p57kip2* and possibly other local maternally expressed imprinted genes. **d)** Imprinting at the Prader-Willi/Angelman syndrome locus is regulated by a bipartite imprinting center (triangles) that confers long-range imprinting control on all the imprinted genes on the two parental alleles. Female germline transmission of the AS-IC is required for methylation and repression of the maternal alleles of the paternally active imprinted genes through inactivation of the PWS-IC. On the paternal chromosome, this AS-IC is nonfunctional, allowing the PWS-IC to confer paternal allelic expression on upstream and downstream genes. *Ube3a* is expressed from the maternal allele in the brain and appears to be associated with an antisense transcript on the paternal allele in a manner similar to the imprinting of *Igf2r*. This antisense transcript is actually the sense transcript for *Snrpn*. In the absence of the paternally inherited PWS-IC, the paternally expressed *Snrpn* gene is repressed, and the paternal allele expresses *Ube3a*. *SnrpN* is also the host transcript for several small nucleolar RNAs that are processed from introns of this 460-kb primary transcript.

how DNA methylation can be important for gene activity through prevention of re-pressor binding.

Consistent with a common regulatory mechanism for IGF2-H19 imprinting, H19-dependent loss of IGF2 imprinting has been described in Wilms tumor [35] and in BWS [36]. However, in hepatoblastoma and in a sizeable proportion of BWS patients, loss of IGF2 imprinting was observed without alteration in H19 imprint-ing, implying H19-independent pathways to loss of imprinting [37]. The DMR1 data suggest that mutations in DMR1, or deficiencies in trans-acting repressor molecules that bind the unmethylated DMR1, could lead to loss of imprinting, with consider-able impact in disease and cancer.

Upstream of *Igf2-H19* is the mouse insulin gene (*ins-2*). This gene is expressed at very low levels in the yolk sac and embryo until late in gestation. This early expres-sion is from the paternal allele only, and *ins-2* appears to be an "innocent bystander" coming under control of the downstream *Igf2* and *H19* enhancers. Later in gesta-tion, insulin becomes up-regulated in the pancreas (its main site of expression) and under the control of its own enhancer. This expression is not imprinted [38].

The *ins-Igf2-H19* domain represents only a fraction of the total imprinted region on mouse distal chromosome 7. Several other imprinted genes are located further upstream and have been the subject of some scrutiny, due to the fact that two of these genes have been implicated in BWS. The first gene, *p57KIP2*, encodes a cyclin-dependent kinase inhibitor that is a negative regulator of the cell cycle. This gene is expressed from the maternally inherited allele. Point mutations have been identified in several BWS patients who inherited the disease from their mothers; these patients do not make a normal gene product [39]. This is consistent with the overgrowth phe-notype seen in BWS patients. The second gene, *KVLQT1*, is very large and, like *p57KIP2*, is expressed from the maternal allele. In BWS patients with chromosomal translocations, the breakpoints map to this gene [40]. *KVLQT1* has a differentially methylated CpG island within the gene that acts as a promoter for a paternally ex-pressed antisense transcript. The DMR comes in methylated from the egg and hence is associated with the active *KVLQT1* allele and the inactive antisense transcript (Fig. 5.2c). Absence of this transcript on the paternal allele results in loss of the antisense transcript and activation of the usually silent paternal *Kvlqt1* allele. This is associated with silencing of the neighboring *p57Kip2* [41], as well as loss of imprinting of sev-eral other genes in the cluster. Hence, this CpG island is an ICR for several im-printed genes but not for the downstream *Igf2-H19* locus. It remains to be shown whether the antisense transcript functions in a manner related to the *Igf2r* antisense transcript described above.

5.2.6
The Mechanisms of Imprinting at the PWS/AS Imprinted Domain (Fig. 5.2d)

Another large imprinted cluster is located on central chromosome 7/human chro-mosome 15q and is associated with PWS and AS. This cluster contains at least 6 im-printed genes and clusters of small nucleolar RNAs (SnoRNAs) within a 3–4-Mb do-main. The region exhibits maternal hypermethylation associated with silencing of the paternally repressed alleles on the maternal chromosome. Two of the genes at the 3' end of the cluster are expressed from the maternally inherited chromosome, and the rest are expressed from the paternally inherited chromosome [18]. Analysis of samples from deletion patients in whom loss of imprinting occurred has defined an imprinting-control region containing two elements and hence thought to act in a bipartite fashion. These imprinting centers (ICs) are (i) the AS-IC that defines the smallest region of overlap in Angelman syndrome patients with deletions and (ii) the PWS-IC that similarly is characterized by the smallest region of overlap in Prader-Willi deletion patients [19].

The bipartite cis-acting imprinting center confers long-range imprinting control on the two parental alleles. Female germline transmission of the AS-IC is required for methylation and repression of the maternal alleles, perhaps through inactivation of the PWS-IC. Maternal deletion of the AS-IC, located considerably upstream of the PWS-IC, results in loss of methylation and activation of the genes that are usually silenced on the maternal chromosome. This causes Angelman syndrome. On the paternal chromosome, the AS-IC is nonfunctional, allowing the PWS-IC to confer paternal allelic expression on upstream and downstream genes. Deletion of the PWS-IC on the paternal chromosome causes methylation and silencing of the paternally expressed genes and activation of the downstream silent paternal allele of *Ube3a* [42]. In humans this causes Prader-Willi syndrome. The PWS-IC is a DMR, and in mouse this allele-specific methylation is found in the egg but not in the sperm. In humans, the maternal-specific methylation is established after fertilization. In mouse, deletion of the PWS-IC on the methylated maternal chromosome has no effect on imprinting, suggesting that the imprinting signal is the unmethylated paternal PWS-IC required to prevent paternal methylation. Although the PWS-IC is known to be a DMR, the AS-IC is not characterized at the molecular level – its precise location and methylation status are not known.

The DMR at the PWS-IC regulates the methylation and silencing of *SNRPN* and both upstream and downstream genes on the maternal allele. Although we do not know how the upstream genes are regulated by this downstream DMR, it appears that the very long SNRPN transcript itself has three functions. First, it encodes the SNRPN/SmN gene products. Second, it acts as the primary host transcript from which the SnoRNAs are processed. Third, its expression from the paternal allele is required for silencing the downstream *UBE3A* gene on the paternal allele [43]. Imprinted silencing of the SNRPN transcript is therefore associated with activation of the *UBE3A* gene on the maternal allele. These results are consistent with a model in which the PWS-IC mediates activation and maintenance of paternal gene expression in the 15q11-q13 region, with repression of the paternal *UBE3A* gene occurring as an indirect result of expression of the SNRPN transcript, which is an antisense transcript of *UBE3A* (Fig. 5.2c). In this model the AS-IC on the maternal chromosome is required for methylation and silencing of those alleles normally paternally expressed. In many respects the relationship of the paternally expressed SNRPN transcript to the reciprocally imprinted maternally expressed *UBE3A* gene is reminiscent of the imprinted antisense transcripts described for KVLQT1 and *Igf2r*.

Additional imprinted domains containing multiple imprinted transcripts have been identified more recently. These include the Gnas/Nesp locus on distal mouse chromosome 2/human chromosome 20q [44] and a large imprinted cluster on mouse distal chromosome 12/human chromosome 14q. The latter cluster has features in common with both the BWS cluster and PWS/AS clusters, and imprinting defects in this region result in both growth and neurological disorders. In addition, comparative analysis has identified genomic and epigenetic features in common with the BWS and PWS/AS loci, including a paternal germline DMR and clusters of SnoRNAs [45, 46]. More detailed analysis of the chromosome-12 cluster therefore may contribute to the search for common mechanisms of imprinting control.

5.3
DNA Methylation and Chromatin

As more experimental reagents have become available for investigating modifications to core histones, an increasing body of evidence proves that DNA methylation is not the only epigenetic modification that affects gene expression. The methyl moieties on DNA serve as a means of differentially marking the two parental alleles and are likely to directly influence chromatin structure and gene expression. Likewise, one cannot ignore the increasing body of evidence suggesting that DNA methylation may be conferred as a result of heritable modifications to the histone proteins required for packaging the DNA into chromatin [47, 48]. So, are germline DMRs a cause or an effect of the chromatin state? What came first, DNA methylation or chromatin modification?

In addition to gene expression, modifications (or lack thereof) of core histones, usually at lysine residues, affect functional features of chromosome architecture, such as heterochromatinization and centromere function, in both vertebrate and invertebrate organisms [2]. Most of our knowledge of the role of histone modifications in gene expression comes from the study of histone acetylation and, more recently, histone methylation. In general, histone acetylation is associated with gene activity, and lack of acetylation, with gene repression. Histone methylation is associated with activity or repression, depending on which lysine is modified. The scope of this chapter doesn't allow justice to be done to the growing literature on histone modification; however, the little we know about the relationships between histone modification and DNA methylation does merit a mention.

In general, DNA methylation is associated with silent CpG island promoters. Gene repression is likely to be mediated by the binding of proteins specific to methylated DNA, the methyl-binding proteins. Several such proteins have been identified; however, their precise functions and target specificity (if any) remain to be elucidated. Methyl-DNA-binding proteins can be found in complexes with histone deacetylases, enzymes that catalyze the deacetylation of histones [49]. These findings suggest that DNA methylation can confer histone deactylation. However, we cannot rule out that deacetylation promotes DNA methylation. Further evidence for a relationship between the two modification types comes from studies on cell lines in which methylation inhibitors and deacetylation inhibitors alone do not alleviate gene silencing although both together can activate genes [50].

Considerable evidence indicates that the chromatin associated with the silent allele of an imprinted domain is more compact than the chromatin of the active allele. For example, in the mouse *Igf2-H19* locus, the differentially methylated *H19* upstream region is in an open chromatin conformation on the active allele, but the silent allele is more resistant to nucleases. The region is also hyperacetylated at histone H4 on the active allele, with the silent allele exhibiting very low levels of acetylation. However, this is not always true. The *Igf2* promoters are not differentially methylated, and both alleles are in an active chromatin conformation. In contrast, these promoters are differentially acetylated in histone H4. Thus, differential histone acetylation can occur in the absence of differential methylation [51]. Less is known about histone methylation at imprinted domains. One report has shown that the maternally inherited PWS-IC is methylated at lysine 9 of H3, with lysine 4 of H3 being methylated on the paternal allele [61]. These finding are consistent with the differential DNA methylation at the PWS-IC and with studies linking silencing with H3 lysine-9 methylation and activity with H3 lysine-4 methylation [52]. One of the real challenges over the next few years will be to uncover the temporal relationships between DNA and chromatin modifications and genome function and, in particular, to determine the situations where DNA methylation regulates chromatin modification and vice versa. Some insight has been gleaned from the studies of that other major mammalian example of gene inactivation – X inactivation.

5.4
X Inactivation

Epigenetic analysis of the silencing of one of the X chromosomes in female mammals, the process of X inactivation, has been extensively reviewed [27]. This process involves many of the regulatory features discussed above, which is not surprising given that, like genomic imprinting, the inactive X is a stable heritable epigenetic state in somatic cells. Indeed, in mouse, X inactivation is imprinted in extra-embryonic tissues, with the paternally inherited copy being preferentially inactivated. X inactivation occurs after fertilization and, in embryonic derivatives, is random, with either of the two X chromosomes being selected for silencing. The inactive X is characterized by asynchronous DNA replication and by epigenetic modifications to the DNA and chromatin, including DNA methylation, histone H3/H4 hypoacetylation, and incorporation of the variant histone macroH2A. Expression of the gene *Xist* from the inactive X is required for initiating X inactivation, and this RNA appears to coat the inactive chromosome. *Xist* is silent on the active X, where an antisense transcript, *Tsix*, is required for *Xist* inactivity. By analyzing the temporal events associated with X inactivation within differentiating female embryonic stem cells, some insight can be gleaned about the hierarchy of interactions resulting in silencing of the chromosome. These studies have shown that *Xist* RNA expression, accumulation, and coating are the first events, which are followed by H3 lysine-9 methylation and hypoacetylation, and H3 lysine-4 hypomethylation. Subsequently, H4 hypoacetylation, gene silencing, and DNA methylation occur [53, 54].

5.5
Prospects and Patterns

Two big issues in the genomic imprinting field continue to challenge investigators. The first is, why did this process evolve? what are the selective pressures that make it a good idea for the mammalian genome to have only one active allele of a gene? why did this parental-origin-specific method of controlling gene dosage evolve? The most popular theory concerning the evolution of imprinting is the genomic-conflict hypothesis proposed by Moore and Haig in 1991 [55]. They suggested that the paternal genome evolved genomic imprinting to extract the maximum resources from the mother, to produce large, "successful" offspring. Paternal interests do not include conservation of maternal resources, since he is likely to have future offspring with a different female. The female, on the other hand, is likely to reproduce several times and each time with a different male. It is in her interests to minimize the drain on her resources for future pregnancies. She therefore has evolved a response to minimize the paternal drive for large offspring. Consistent with this is the propensity for paternally expressed imprinted genes to be involved in growth enhancement and maternally expressed genes to be negative regulators of growth. In addition many imprinted genes function in the development of the placenta, where one might expect a battle for maternal resources to occur. After birth, other paternally expressed imprinted genes appear to be required for appropriate maternal care of offspring.

However, several findings do not fit neatly into the conflict hypothesis. Many imprinted genes are expressed in neural tissue and do not apparently regulate processes involved in nurture or growth. Other paternally expressed imprinted genes result in placental overgrowth, which is either detrimental to the fetus or has no effect on fetal well-being. Furthermore, Mash2, a maternally expressed imprinted gene, is essential for developing a normal functioning placenta – an attribute (one might predict from the theory) that should be afforded to a paternally expressed imprinted gene. Nonetheless, there is no doubt that imprinted genes contribute to prenatal growth and to the development of particular lineages and that altering the dosage of parental chromosomes or the imprinted genes located there has major implications for normal mammalian development. As more imprinted genes become identified and characterized functionally, the genomic-conflict theory might be refined or other theories may find favor.

The second issue of considerable interest involves the evolution of the imprinting mechanism(s). Are there specific genomic features or sequences that confer imprinting control? Apart from the obvious association with CpG islands, are there common regulatory elements that are acted upon differentially in the male and female germlines and that result in heritable allele-specific modifications that cause genes to be expressed in a parental-origin-specific manner? If so, how and why do the male and female germlines act differentially on these regions?

Most methylated DNA within the genome is located in repetitive parasitic sequence elements such as transposons and endogenous retroviruses, which constitute about a third of the human genome. The primary function of DNA methylation might therefore be as a host-defense mechanism that protects against transcriptional activation of these repetitive elements [56]. 5-methyl cytosine has a tendency to be deaminated to thymidine resulting in a change in the sequence and a drive towards genetic rather than epigenetic inactivation. As human and mouse genome sequences become more amenable to investigation, the relationship, if any, between DNA methylation, endogenous gene silencing, and the silencing of genomic parasitic elements can be addressed more fully. Interestingly, there are two reports of imprinting of retrotransposon-like elements [57, 58]; both elements are paternally expressed and members of a rare family of retrotransposon-like elements in the mammalian genome. Furthermore, an intra-cisternal A sequence, a murine retrotransposon-like element, can confer epigenetic regulation on a neighboring gene [59]. A tandemly repeated element (41mer × 40) also plays a direct role in imprinting the *Rasgrf1*gene. This element is located 30 kb upstream of the promoter and lies adjacent to a DMR. The region is methylated on the active paternal allele and ummethylated on the maternal chromosome. Upon deletion of the paternally inherited tandem repeat, the DMR loses methylation and the gene becomes silent [60]. The precise mechanism involved in this long-range epigenetic control remains to be elucidated; however, the repeat is required for methylation in the male germline. Detailed comparative sequence analysis of imprinted domains both within and between species is likely to tell us about the locations of parasitic and other repetitive elements relative to imprinting-controlling elements and to the imprinted genes themselves. These findings should provide clues as to whether there are common genomic features that confer epigenetic control on adjacent genes.

However, at the moment the search for such recurrent themes in the epigenetic control of gene expression has yielded few common patterns, apart from the importance of CpG islands in the process. Until more is known about the epigenetic control of developmental genes that are biallelically expressed, the study of imprinted and X-linked genes allows direct investigation of the role of DNA and chromatin modification in transcriptional control. Perhaps one recurrent theme that emerges from these studies is the presence of noncoding RNAs that in some, although not all, cases are expressed in an antisense direction on an allele that is normally silent in the sense direction. The *Igf2-H19* locus is likely to be an exception to this and may represent a second class of epigenetic regulation; one involving methylation-sensitive insulators. The speculation that two types of imprinting regulation occur – one involving non-coding RNAs and the other involving methylation-dependent germline-imprinted intergenic insulators – is perhaps premature and can only be answered when we understand more fully the precise nature of regulatory elements in imprinted domains and have conducted detailed comparative analysis of large domains subject to epigenetic control.

References

1 BIRD, A., DNA methylation patterns and epigenetic memory, *Genes Dev.* **2002**, *16*, 6–21.

2 RICHARDS, E. J.; ELGIN, S. C., Epigenetic codes for heterochromatin formation and silencing: rounding up the usual suspects, *Cell* **2002**, *108*, 489–500.

3 MEEHAN, R. R.; STANCHEVA, I., DNA methylation and control of gene expression in vertebrate development, *Essays Biochem.* **2001**, *37*, 59–70.

4 WALSH, C. P.; BESTOR, T. H., Cytosine methylation and mammalian development, *Genes Dev.* **1999**, *13*, 26–34.

5 SHEMER, R.; WALSH, A.; EISENBERG, S.; BRESLOW, J. L.; RAZIN, A., Tissue-specific methylation patterns and expression of the human apolipoprotein AI gene, *J. Biol. Chem.* **1990**, *265*, 1010–1015.

6 FUTSCHER, B. W.; OSHIRO, M. M.; WOZNIAK, R. J.; HOLTAN, N.; HANIGAN, C. L.; DUAN, H.; DOMANN, F. E., Role for DNA methylation in the control of cell type specific maspin expression, *Nat, Genet.* **2002**, *31*, 175–179.

7 JACKSON-GRUSBY, L.; BEARD, C.; POSSEMATO, R.; TUDOR, M.; FAMBROUGH, D.; CSANKOVSZKI, G.; DAUSMAN, J.; LEE, P.; WILSON, C.; LANDER, E.; JAENISCH, R., Loss of genomic methylation causes p53-dependent apoptosis and epigenetic deregulation, *Nature Genet.* **2001**, *27*, 31–39.

8 FERGUSON-SMITH, A. C.; SURANI, M. A., Imprinting and the epigenetic asymmetry between parental genomes, *Science* **2001**, *293*, 1086–1089.

9 BEECHEY, C. **2002**, http://www.mgu.har.mrc.ac.uk/imprinting/all_impmaps.html

10 ENGEMANN, S.; STRODICKE, M.; PAULSEN, M.; FRANCK, O.; REINHARDT, R.; LANE, N.; REIK, W.; WALTER, J., Sequence and functional comparison in the Beckwith-Wiedemann region: implications for a novel imprinting centre and extended imprinting. *Hum. Mol. Genet.* **2000**, *9*, 2691–2706.

11 DECHIARA, T. M.; ROBERTSON, E. J.; EFSTRATIADIS, A., Parental imprinting of the mouse insulin-like growth factor II gene, *Cell* **1991**, *64*, 849–859.

12 FERGUSON-SMITH, A. C.; CATTANACH, B. M.; BARTON, S. C.; BEECHEY, C. V.; SURANI, M. A., Embryological and molecular investigations of parental imprinting on mouse chromosome 7, *Nature* **1991**, *351*, 667–670.

13 CONSTANCIA, M.; HEMBERGER, M.; HUGHES, J.; DEAN, W.; FERGUSON-SMITH, A.; FUNDELE, R.; STEWART, F.; KELSEY, G.; FOWDEN, A.; SIBLEY, C.; REIK, W., Placental-specific IGF-II is a major modulator of placental and fetal growth, *Nature* **2002**, *417*, 945–948.

14 FRANK, D.; FORTINO, W.; CLARK, L.; MUSALO, R.; WANG, W.; SAXENA, A.; LI, C. M.; REIK, W.; LUDWIG, T.; TYCKO, B., Placental overgrowth in mice lacking the imprinted gene Ipl, *Proc Natl Acad Sci USA* **2002**, *99*, 7490–7495.

15 GUILLEMOT, F.; CASPARY, T.; TILGHMAN, S. M.; COPELAND, N. G.; GILBERT, D. J.; JENKINS, N. A.; ANDERSON, D. J.; JOYNER, A. L.; ROSSANT, J.; NAGY, A., Genomic imprinting of Mash2, a mouse gene required for trophoblast development, *Nature Genet.* **1995**, *9*, 235–242.

16 TAKAHASHI, K.; KOBAYASHI, T.; KANAYAMA, N., p57(Kip2) regulates the proper development of labyrinthine and spongiotrophoblasts, *Mol. Hum. Reprod.* **2000**, *6*, 1019–1025.

17 PAULSEN, M.; FERGUSON-SMITH, A. C., DNA methylation in genomic imprinting, development, and disease, *J. Pathol.* **2001**, *195*, 97–110.

18 NICHOLLS, R. D.; KNEPPER, J. L., Genome organization, function, and imprinting in Prader-Willi and Angelman syndromes, *Annu. Rev. Genomics Hum. Genet.* **2001**, *2*, 153–175.

19 CASSIDY, S. B.; DYKENS, E.; WILLIAMS, C. A., Prader-Willi and Angelman syndromes: sister imprinted disorders, *Amer. J. Med. Genet.* **2000**, *97*, 136–146.

20 BARLOW, D. P.; STOGER, R.; HERRMANN, B. G.; SAITO, K.; SCHWEIFER, N., The mouse insulin-like growth factor type-2 receptor is imprinted and closely linked

to the Tme locus, *Nature* **1991**, *349*, 84–87.

21 RIESEWIJK, A. M.; SCHEPENS, M. T.; WELCH, T. R.; VAN DEN BERG-LOONEN, E. M.; MARIMAN, E. M.; ROPERS, H. H.; KALSCHEUER, V. M., Maternal-specific methylation of the human IGF2R gene is not accompanied by allele-specific transcription, *Genomics* **1996**, *31*, 158–166.

22 LI, E.; BEARD, C.; JAENISCH, R., Role for DNA methylation in genomic imprinting, *Nature* **1993**, *366*, 362–365.

23 TANAKA, M.; PUCHYR, M.; GERTSENSTEIN, M.; HARPAL, K.; JAENISCH, R.; ROSSANT, J.; NAGY, A., Parental origin-specific expression of Mash2 is established at the time of implantation with its imprinting mechanism highly resistant to genome-wide demethylation, *Mech. Dev.* **1999**, *87*, 129.

24 LUDWIG, T.; EGGENSCHWILER, J.; FISHER, P.; D'ERCOLE, A. J.; DAVENPORT, M. L.; EFSTRATIADIS, A., Mouse mutants lacking the type 2 IGF receptor (IGF2R) are rescued from perinatal lethality in Igf2 and Igf1r null backgrounds, *Dev. Biol.* **1996**, *177*, 517–535.

25 STOGER, R.; KUBICKA, P.; LIU, C. G.; KAFRI, T.; RAZIN, A.; CEDAR, H.; BARLOW, D. P., Maternal-specific methylation of the imprinted mouse Igf2r locus identifies the expressed locus as carrying the imprinting signal, *Cell* **1993**, *73*, 61–71.

26 SLEUTELS, F.; ZWART, R.; BARLOW, D. P., The non-coding Air RNA is required for silencing autosomal imprinted genes, *Nature* **2002**, *415*, 810–813.

27 AVNER, P.; HEARD, E., X-chromosome inactivation: counting, choice and initiation, *Nature Rev. Genet.* **2001**, *2*, 59–67.

28 GRANDJEAN, V.; SMITH, J.; SCHOFIELD, P. N.; FERGUSON-SMITH, A. C., Increased IGF-II protein affects p57kip2 expression in vivo and in vitro: implications for Beckwith-Wiedemann syndrome, *Proc. Natl. Acad. Sci. USA* **2000**, *97*, 5279–5284.

29 LEIGHTON, P. A.; SAAM, J. R.; INGRAM, R. S.; STEWART, C. L.; TILGHMAN, S. M., An enhancer deletion affects both H19 and Igf2 expression, *Genes Dev.* **1995**, *9*, 2079–2089.

30 SASAKI, H.; JONES, P. A.; CHAILLET, J. R.; FERGUSON-SMITH, A. C.; BARTON, S. C.; REIK, W.; SURANI, M. A., Parental imprinting: potentially active chromatin of the repressed maternal allele of the mouse insulin-like growth factor II (Igf2) gene, *Genes Dev.* **1992**, *6*, 1843–1856.

31 FERGUSON-SMITH, A. C.; SASAKI, H.; CATTANACH, B. M.; SURANI, M. A., Parental-origin-specific epigenetic modification of the mouse H19 gene, *Nature* **1993**, *362*, 751–755.

32 CONSTANCIA, M.; DEAN, W.; LOPES, S.; MOORE, T.; KELSEY, G.; REIK, W., Deletion of a silencer element in Igf2 results in loss of imprinting independent of H19, *Nature Genet.* **2000**, *26*, 203–206.

33 FERGUSON-SMITH, A. C., Genetic imprinting: silencing elements have their say, *Curr. Biol.* **2000**, *10*, R872–875.

34 EDEN, S.; CONSTANCIA, M.; HASHIMSHONY, T.; DEAN, W.; GOLDSTEIN, B.; JOHNSON, A. C.; KESHET, I.; REIK, W.; CEDAR, H., An upstream repressor element plays a role in Igf2 imprinting, *EMBO J.* **2001**, *20*, 3518–3525.

35 FEINBERG, A. P., Imprinting of a genomic domain of 11p15 and loss of imprinting in cancer: an introduction, *Cancer Res.* **1999**, *59*(7 Suppl), 1743s–1746s.

36 ENGEL, J. R.; SMALLWOOD, A.; HARPER, A.; HIGGINS, M. J.; OSHIMURA, M.; REIK, W.; SCHOFIELD, P. N.; MAHER, E., Epigenotype-phenotype correlations in Beckwith-Wiedemann syndrome, *J. Med. Genet.* **2000**, *37*, 921–926.

37 JOYCE, J. A.; LAM, W. K.; CATCHPOOLE, D. J.; JENKS, P.; REIK, W.; MAHER, E. R.; SCHOFIELD, P. N., Imprinting of IGF2 and H19: lack of reciprocity in sporadic Beckwith-Wiedemann syndrome, *Hum. Mol. Genet.* **1997**, *6*, 1543–1548.

38 GIDDINGS, S. J.; KING, C. D.; HARMAN, K. W.; FLOOD, J. F.; CARNAGHI, L. R., Allele specific inactivation of insulin 1 and 2, in the mouse yolk sac, indicates imprinting, *Nature Genet.* **1994**, *6*, 310–313.

39 HATADA, I.; OHASHI, H.; FUKUSHIMA, Y.; KANEKO, Y.; INOUE, M.; KOMOTO, Y.; OKADA, A.; OHISHI, S.; NABETANI, A.; MORISAKI, H.; NAKAYAMA, M.;

NIIKAWA, N.; MUKAI, T., An imprinted gene p57KIP2 is mutated in Beckwith-Wiedemann syndrome, *Nature Genet.* **1996**, *14*, 171–173.

40 LEE, M. P.; HU, R. J.; JOHNSON, L. A.; FEINBERG, A. P., Human KVLQT1 gene shows tissue-specific imprinting and encompasses Beckwith-Wiedemann syndrome chromosomal rearrangements, *Nature Genet.* **1997**, *15*, 181–185.

41 HORIKE, S.; MITSUYA, K.; MEGURO, M.; KOTOBUKI, N.; KASHIWAGI, A.; NOTSU, T.; SCHULZ, T. C.; SHIRAYOSHI, Y.; OSHIMURA, M., Targeted disruption of the human LIT1 locus defines a putative imprinting control element playing an essential role in Beckwith-Wiedemann syndrome, *Hum. Mol. Genet.* **2000**, *9*, 2075–2083.

42 CHAMBERLAIN, S. J.; BRANNAN, C. I., The Prader-Willi syndrome imprinting center activates the paternally expressed murine Ube3a antisense transcript but represses paternal Ube3a, *Genomics* **2001**, *73*, 316–322.

43 RUNTE, M.; HUTTENHOFER, A.; GROSS, S.; KIEFMANN, M.; HORSTHEMKE, B.; BUITING, K., The IC-SNURF-SNRPN transcript serves as a host for multiple small nucleolar RNA species and as an antisense RNA for UBE3A, *Hum. Mol. Genet.* **2001**, *10*, 2687–2700.

44 WROE, S. F.; KELSEY, G.; SKINNER, J. A.; BODLE, D.; BALL, S. T.; BEECHEY, C. V.; PETERS, J.; WILLIAMSON, C. M., An imprinted transcript, antisense to Nesp, adds complexity to the cluster of imprinted genes at the mouse Gnas locus, *Proc. Natl. Acad. Sci. USA* **2000**, *97*, 3342–3346.

45 TAKADA, S.; PAULSEN, M.; TEVENDALE, M.; TSAI, C. E.; KELSEY, G.; CATTANACH, B. M.; FERGUSON-SMITH, A. C., Epigenetic analysis of the Dlk1-Gtl2 imprinted domain on mouse chromosome 12: implications for imprinting control from comparison with Igf2-H19, *Hum. Mol. Genet.* **2002**, *11*, 77–86.

46 CAVAILLE, J.; SEITZ, H.; PAULSEN, M.; FERGUSON-SMITH, A. C.; BACHELLERIE, J. P., Identification of tandemly-repeated C/D snoRNA genes at the imprinted human 14q32 domain reminis-

cent of those at the Prader-Willi/Angelman syndrome region, *Hum. Mol. Genet.* **2002**, *11*, 1527–1538.

47 TAMARU, H.; SELKER, E. U., A histone H3 methyltransferase controls DNA methylation in *Neurospora crassa*, *Nature* **2001**, *414*, 277–283.

48 JACKSON, J. P.; LINDROTH, A. M.; CAO, X.; JACOBSEN, S. E., Control of CpNpG DNA methylation by the KRYPTONITE histone H3 methyltransferase, *Nature* **2002**, *416*, 556–560.

49 HENDRICH, B.; BIRD, A., Mammalian methyltransferases and methyl-CpG-binding domains: proteins involved in DNA methylation, *Curr. Top. Microbiol. Immunol.* **2000**, *249*, 55–74.

50 CAMERON, E. E.; BACHMAN, K. E.; MYOHANEN, S.; HERMAN, J. G.; BAYLIN, S. B., Synergy of demethylation and histone deacetylase inhibition in the re-expression of genes silenced in cancer, *Nature Genet.* **1999**, *21*, 103–107.

51 GRANDJEAN. V.; O'NEILL, L.; SADO, T.; TURNER, B.; FERGUSON-SMITH, A., Relationship between DNA methylation, histone H4 acetylation and gene expression in the mouse imprinted Igf2-H19 domain, *FEBS Lett.* **2001**, *488*, 165–169.

52 XIN, Z.; ALLIS, C. D.; WAGSTAFF, J., Parent-specific complementary patterns of histone H3 lysine 9 and H3 lysine 4 methylation at the Prader-Willi syndrome imprinting center, *Amer. J. Hum. Genet.* **2001**, *69*, 1389–1394.

53 HEARD, E.; ROUGEULLE, C.; ARNAUD, D.; AVNER, P.; ALLIS, C. D.; SPECTOR, D. L., Methylation of histone H3 at Lys-9 is an early mark on the X chromosome during X inactivation, *Cell* **2001**, *107*, 727–738.

54 MERMOUD, J. E.; POPOVA, B.; PETERS, A. H.; JENUWEIN, T.; BROCKDORFF, N., Histone H3 lysine 9 methylation occurs rapidly at the onset of random X chromosome inactivation, *Curr Biol.* **2002**, *12*, 247–251.

55 MOORE, T.; HAIG, D., Genomic imprinting in mammalian development: a parental tug-of-war, *Trends Genet.* **1991**, *7*, 45–49.

56 YODER, J. A.; WALSH, C. P.; BESTOR, T. H., Cytosine methylation and the ecology of intragenomic parasites, *Trends Genet.* **1997**, *13*, 335–340.

57 ONO, R.; KOBAYASHI, S.; WAGATSU-MA, H.; AISAKA, K.; KOHDA, T.; KANEKO-ISHINO, T.; ISHINO, F., A retrotransposon-derived gene, PEG10, is a novel imprinted gene located on human chromosome 7q21, *Genomics* **2001**, *73*, 232–237.

58 CHARLIER, C.; SEGERS, K.; WAGEN-AAR, D.; KARIM, L.; BERGHMANS, S.; JAILLON, O.; SHAY, T.; WEISSENBACH, J.; COCKETT, N.; GYAPAY, G.; GEORGES, M., Human-ovine comparative sequencing of a 250-kb imprinted domain encompassing the callipyge (clpg) locus and identification of six imprinted transcripts: DLK1, DAT, GTL2, PEG11, anti-PEG11, and MEG8, *Genome Res.* **2001**, *11*, 850–862.

59 MORGAN, H. D.; SUTHERLAND, H. G.; MARTIN, D. I.; WHITELAW, E., Epigenetic inheritance at the agouti locus in the mouse, *Nature Genet.* **1999**, *23*, 314–318.

60 YOON, B. J.; HERMAN, H.; SIKORA, A.; SMITH, L. T.; PLASS, C.; SOLOWAY, P. D., Regulation of DNA methylation of Rasgrf1, *Nature Genet.* **2002**, *30*, 92–96.

61 XIN, Z.; ALLIS, C. D.; WAGSTAFF, J., Parent-specific complementary patterns of histone H3 lysine 9 and H3 lysine 4 methylation at the Prader-Willi syndrome imprinting center, *Am. J. Hum. Gen.* **2001**, *69*, 1389–1394.

6
Epigenetic Trouble:
Human Diseases Caused by Epimutations

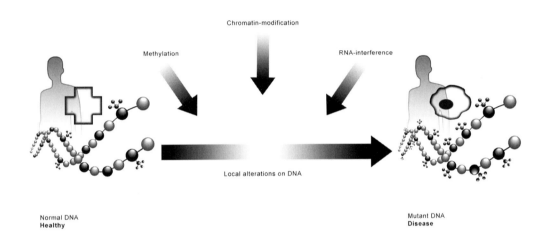

6

Epigenetic Trouble: Human Diseases Caused by Epimutations

Jörn Walter* and Martina Paulsen

Summary

Epigenetic modifications are becoming increasingly important in our understanding of normal and abnormal gene regulation processes. The stable inheritance of open or closed chromatin configurations seems to be a key determinant for differential gene expression in humans. The control of such epigenetic inheritance depends on multiple layers of interacting epigenetic modifications. Several distinct syndromes and diseases are caused by local epigenetic alterations of the chromatin structure. Such epimutations can be caused by incorrect epigenetic reprogramming during development, by mutations in enzymes involved in epigenetic inheritance, or by mutations of cis-acting DNA sequences. It is to be expected that epigenetic variations widely contribute to altered gene expression in numerous complex diseases.

6.1
Introduction

With the completion of the human genome, one of the major future goals is to understand the molecular mechanisms regulating the differential expression of all our genes. In recent years, it has become more and more evident that multiple layers of epigenetic modifications control the expression of most, if not all, human genes. Epigenetic modifications influence the chromatin structure and hence the accessibility of the genetic information to the transcriptional machinery. These modifications are reversible and hence do not change the genetic content of the chromosomes, and they can be stably replicated over generations of cell division. The correct inheritance of epigenetic programs in the genome is necessary for normal development [1]. Epimutations, i.e., incorrectly established epigenetic programs, result in aberrant gene expression and thus contribute to a variety of human disorders [2–5].

Epigenetic modifications are essential for several genetic functions controlling transcriptional activity and for the stability of chromosomes [2, 6–8]. Epigenetic me-

* Corresponding author

chanisms may even operate outside the nucleus, by transmitting signals from one cell to another to affect gene expression in other cells. Examples of such sophisticated ways of control are found in plants and fungi and are mediated through small double-stranded RNA molecules [9, 10]. The main areas of epigenetic control, however, are the chromosomes, which are controlled by several layers of epigenetic changes: modifications of histones in association with heterochromatin proteins and noncoding RNAs, and modifications of the DNA [2, 6, 11]. Whereas chromatin modifications are found in all eukaryotes, DNA modifications such as DNA methylation are of particular epigenetic importance in mammals and humans. Here, DNA modifications are confined to cytosines in the dinucleotide CpG. DNA methylation, however, is intimately linked to chromatin structures and chromatin-modifying enzymes [6].

Although the complexity of such diverse epigenetic codes and their functions in controlling gene expression are far from being understood, many human diseases are causally linked to epimutations, a term coined by R. Holliday [12]. Epigenetic diseases include cancer (reviewed in [3], see also Millar et al., Chapter 1), aspects of aging (see Issa, Chapter 8), imprinting syndromes, immunological diseases, and developmental disorders of the CNS. In this review we focus on examples in which DNA-methylation changes play a central role in disease development. The many studies on the molecular origins of such diseases and disorders have substantially contributed to our understanding of the complexity of epigenetic mechanisms. They also nicely demonstrate the intimate link between the different layers of epigenetic regulation and their influence on gene control and gene function.

6.2
Mechanisms of DNA Methylation

A key epigenetic modification in humans is methylation of the carbon-5 of cytosines in DNA. Substantial progress has been made recently in understanding the key players controlling DNA methylation. DNA methylation is catalyzed by several independently operating DNA methyltransferases (DNMTs) [13]. In humans and mammals, the key enzymes are DNMT1, DNMT3A, and DNMT3B. All three methyltransferases are essential for development and viability, since knockout mutations in any of these genes lead to embryonic or postnatal death in mice [14, 15]. DNMT1 is the first and best characterized methyltransferase. In vitro, it can establish methylation patterns de novo [16]. In vivo, DNMT1 is associated with the replication machinery and seems to be a key enzyme for maintaining methylation patterns on a wide range of single-copy sequences [17–19]. Dnmt3a/DNMT3A and Dnmt3b/DNMT3B of mouse and human have been identified on the basis of sequence homology and were initially characterized as de novo methyltransferases with preferential de novo methylation activity of the in vitro-purified enzymes [15, 20]. In vivo, the loss of these enzymes leads to a dramatic reduction in the methylation of repetitive DNA sequences [15, 21]. In addition, recent studies demonstrate that they are capable of de-novo methylating sequences that are usually recognized by DNMT1 and thus compensate for inefficient maintenance methylation by DNMT1 [22]. The loss of DNMT1 and DNMT3B activity in (some) cancer cell lines decreases the overall level of methylation to less than 5%. Hence, all three methyltransferases are structurally and functionally distinct enzymes with partially overlapping and/or redundant functions [23]. The N-terminal regions of all three methyltransferases contain distinct domains interacting with a range of different proteins or protein complexes, including replication complexes, transcriptional co-repressors and histone deacetylases [6, 19, 24]. The distinct mechanisms and functions of all three methyltransferases, i.e., their ability to de-novo methylate or maintain methylation at certain sequences, is mediated through differential interaction of the methyltransferases with other proteins.

However, how methyltransferases establish specific methylation patterns along the chromosomes remains an open question. Several scenarios can be envisaged: In the first, methyltransferases bind to co-repressors or DNA-binding protein complexes, which guide the enzymes to specific chromosomal regions, inducing local hypermethylation and deacetylation (Fig. 6.1A). Examples of such a scenario have been described: DNMT1 is found in complexes with the RB (retinoblastoma) gene product, and such complexes also include the transcription factor E2F1 and the histone deacetylase HDAC1 [24, 25]. DNMT1 also interacts with a replication-associated protein complex formed by DMAP1 (DNMT1-associated protein) TG101 (a transcriptional co-repressor), and HDAC2 (histone deacetylase 2) [19]. The murine Dnmt3a complexes with RP58, a DNA-binding transcriptional regulator, and HDAC1. All three proteins colocalize on transcriptionally silent heterochromatin [26].

A second possibility is that methyltransferases have a widespread ability to induce or maintain methylation and that specific proteins or chromatin structures selectively protect certain regions against enzymatic accessibility.

A third possibility is that such widespread methylation activity of methyltransferases is locally antagonized by the enzymatic action of DNA demethylases. The frequently observed mosaic methylation patterns of many genes support this hypothesis [27] (see also Reik and Dean, Chapter 4).

A fourth possibility is that certain DNA sequences or secondary DNA structures are specifically targeted by certain DNMTs. In vitro, DNMT1 shows a high affinity for hairpin structures in repetitive sequences, and DNMT3A and B are important in maintaining methylation in repetitive sequences [8, 15].

In summary, the precise establishment and maintenance of specific methylation patterns is highly regulated and depends on the specific interaction of distinct methyltransferases with cis-acting (DNA structures) and trans-acting (associated proteins) factors. Mutations in DNMTs, their interacting partners, or DNA sequences have a profound influence on specific DNA methylation patterns and, consequently, on the generation of disease phenotypes.

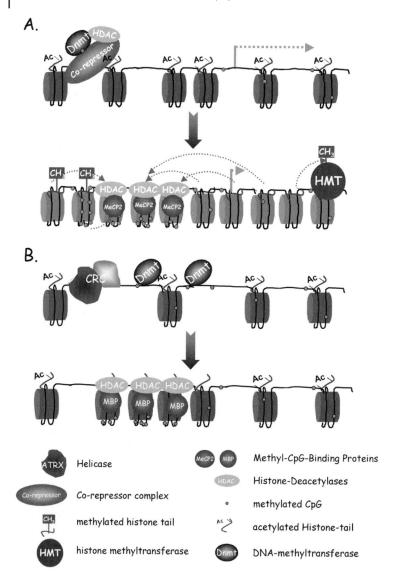

Fig. 6.1 Molecular recognition of methylation patterns. The figure presents models for chromatin alterations induced by combined effects of DNA-methylation, histone modifications and nucleosome remodelling. The diagrams show the DNA (line) wound around nucleosomes (cylinders). The location and activity of hypothetical genes in the vicinity of epigenetic changes is indicated by a dotted line (arrow = active gene; circle = silenced gene). A description of the various enzymes and modifications is given below. **A.** DNMTs are targeted to a chromosomal region, inducing methylation followed by binding of the methyl-binding-proteins (MeCP2 is shown as an example) and deacetylation of histones by HDACs and methylation of histones by histone-methyltransferases (HMTs). **B.** Remodelling of the nucleosomal structure (compaction) by chromatin remodelling complexes (ATRX) induces regional *de novo* methylation by DNMTs followed by binding of methyl binding proteins (MBPs) and chromatin condensation.

6.3
Molecular Recognition of Methylation Patterns

DNA methylation has been widely discussed in the past 20 years as being influential in gene expression [28, 29]. However, the question of how DNA-methylation patterns are associated with inactive chromatin structures remained open for a long time. Recently, some of the molecular details have been discovered, showing an intimate connection between DNMTs/DNA-methylation patterns and HDACs, HMT/Histone-modifications controlling chromatin structure [2, 6, 30]. Several such connecting pathways have been described.

In the first pathway, the translation of DNA methylation into altered chromatin structure is mediated by direct interaction of methyltransferases with histone deacetylases (HDACs) (Fig. 6.1B). This proposal was based on the observation that DNMT1 coimmunoprecipitates with HDAC1 [19, 24, 26, 31].

The key players in the second pathway are methyl-binding proteins (MBPs) that specifically recognize methylated CpGs and bind to them via methyl-binding domains (MBDs). MBPs such as MeCP2 or MBD1–4 function as adaptors by connecting chromatin-modifying complexes (CMCs) to methylated regions of the genome [32–36] (Fig. 6.1 A and B). The composition of these CMCs differs and may include transcriptional repressors (MBD1, Sin3A, RbAp46/48), histone deacetylases (HDAC1 and 2), histone methyltransferases (HMTs), and chromatin-remodeling helicases (SWI2/SNF2). Mutations in such helicases lead to dramatic loss of DNA methylation in plants and animals [37, 38].

Although it seems clear that MBPs and helicases modulate the epigenetic codes by translating DNA-methylation patterns into differentially packaged chromatin, the consequence of events might not be as outlined above. Recent studies in plants and fungi indicate a reverse scenario of interaction between DNA methylation and chromatin modifications: mutations in the *dim2* gene, a histone methyltransferase in the fungus *Neurospora crassa*, lead to a complete loss of DNA methylation, indicating that the establishment and maintenance of DNA methylation depends on histone methylation and chromatin alterations [37, 39]. Analysis of histone modifications (in the form of differential acetylation and methylation) on the inactive X chromosome in mouse seems to support this order of events. Here, both deacetylation of histone H4 and methylation of histone H3 at particular residues are one of the earliest modifications of the inactive X chromosome and occur before DNA methylation [40, 41]. Overall, it seems most likely that both sequences of events are likely to occur in vivo and that the sequence of events is largely dependent on the chromosomal context.

Groundbreaking studies on MBDs and their interactions with histone-modifying complexes have elucidated a complex network of communication between different layers of epigenetic modifications of the chromatin and the DNA [2, 6, 30]. Moreover, they have demonstrated that changes in the interaction of these epigenetic codes, caused by mutations in methyltransferases, MBPs, or chromatin-modifying proteins, have profound effects on the developmental potential of the organism and on the generation of diseases. Important examples of such epigenetic defects are mutations in the methyltransferase DNMT3B, the MBP MeCP2, and the proposed chromatin-remodeling enzyme ATRX, all of which cause distinct disease symptoms intimately linked to alterations in specific DNA-methylation patterns. Other examples of epigenetic diseases originate from genetic alterations like tri-nucleotide expansions, as in fragile X-syndrome, or deletions/translocations, as in some imprinting syndromes (see below). Imprinting syndromes can, however, also arise on a purely epigenetic basis, i.e., without any genetic alteration. Such epimutations are most likely caused by aberrant epigenetic reprogramming during early development ([1], Reik and Dean, Chapter 4).

6.4
Rett Syndrome

Several studies have shown that DNA methylation is crucial for normal function and development of neurons in the brain [42, 43]. Mutations in an X-linked gene encoding the MeCP2 protein, which binds to methylated DNA and links DNA methylation to transcriptional repression, cause a severe neurological disorder called Rett syndrome (RTT, MIM 312750) [32, 44]. Rett syndrome is the most common sporadically inherited form of mental retardation in females, with an incidence of 1 in 10–15 000. The earliest neurological symptoms occur between 6 and 18 months after birth and are followed by phases of stabilization and disease progression. The neurological defects lead to a range of mental problems: degeneration of speech and acquired motor skills, seizures, autism, ataxia, and most characteristically, stereotypical hand movements. The missense and truncating mutations in the *MeCP2* gene found in Rett patients are scattered all over the gene. There seems to be a strict correlation between the location, the form of the mutation and the severity of the symptoms [45, 46]. Several mutations are clustered in two functionally characterized domains of the protein: the methyl-CpG binding domain (MBD) and the transcriptional repression domain (TRD), through which interaction with other protein complexes occurs. Biochemical characterization of the mutations in the MBD have shown disturbed binding to methylated CpGs, mostly caused by protein misfolding [47, 48]. Knockout mutations of the mouse Mecp2 reproduce the key aspects of this disease, although the symptoms are less severe. Mice with conditional knockouts of the *MeCP2* gene in the brain show neuronal degeneration and behavioral symptoms resembling those of Rett syndrome in humans [49, 50]. These studies also reveal that the effects of MeCP2 loss occur only in postmitotic neurons. MeCP2 is associated with the inactive, methylated X chromosome. A methylation study on two X-chromosomal genes in cells of Rett-syndrome patients suggests a significant decrease in methylation of one of the genes [51].

In summary, the defects found in Rett patients seem to be caused by reduced MeCP2 function in neuronal cells in the brain. Why such specific pathological effects are confined to postmitotic neuronal cells remains an open question, since MeCP2 is ubiquitously expressed in many cell types during development.

6.5
ICF Syndrome

The ICF syndrome is a rare autosomal-recessive disease characterized by (variable) immunodeficiency, instability of the centromeric region, and facial anomalies (ICF, MIM 242806). On the molecular level, the syndrome is accompanied by hypomethylation of the classical satellite 2 and 3 sequences [52, 53]. The ICF gene was mapped to 20q11-q13, a region that contains the DNMT3B gene [54]. Several groups then independently showed that the disease is indeed caused by mutations in the DNA methyltransferase 3B gene (DNMT3B) [15, 54, 55]. Most ICF patients have reductions or absence of at least two immunoglobulin isotypes, causing defective cell-mediated immunity. In addition, certain chromosomes (chromosomes 1, 9, and 16) in mitogen-stimulated lymphocytes of ICF patients exhibit pericentromeric condensation anomalies and enhanced chromosomal instability [56]. These alterations coincide with a mild decrease in overall genomic 5-methyl cytosine levels and a striking hypomethylation of repetitive sequences in the pericentromeric region of these chromosomes [57]. Microarray expression analysis shows a small number of genes with altered expression levels in lymphocytes of ICF patients [58]. Half of these genes play a role in controlling the immune function. However, no changes in promoter methylation were observed in those genes, indicating that the DNMT3B mutations in ICF syndrome cause lymphogenesis-associated gene dysregulation by indirect effects on gene expression. Mutations causing a severe form of ICF are mainly clustered in the characteristic conserved motifs of methyltransferases in the catalytic domain of DNMT3B. Knockout mutations of the homologous *Dnmt3b* gene in the mouse show some similarities to the molecular phenotypes observed in ICF patients, in that repetitive sequences are hypomethylated [15]. However, whereas point mutations in the human do not appear to be lethal, mice lacking the *Dnmt3b* gene die after 9.5–10.5 days of embryonic development [15].

In summary, the observed chromatin alterations and instabilities in ICF patients seem to be directly correlated with the loss of DNA methylation caused by a reduced function of the DNMT3B methyltransferase activity. However, independent of its function as a methyltransferase, DNMT3B may also operate as a transcriptional corepressor [59]. This may explain the differences between the ICF phenotypes in humans, caused by point mutations mainly in the catalytic methyltransferase domain, and the lethal phenotype of null mutants in the mouse.

6.6
ATR-X Syndrome

A third disease showing abnormal DNA-methylation patterns is the ATR-X (X-linked alpha-thalassemia/mental retardation) syndrome (MIM 301040). The *ATRX* gene codes for a protein that contains a PHD-like domain (PHD = plant homeodomain), which is present in many chromatin-associated proteins. In addition, the protein contains a carboxy-terminal domain, which identifies it as a member of the SNF2 family of helicase/ATPases [60].

Mutations in the *ATRX* gene give rise to characteristic developmental abnormalities, including severe mental retardation, facial dysmorphism, urogenital abnormalities, alpha-thalassaemia, and sex-reversal phenotypes [61, 62]. The ATRX protein is supposed to act as a transcriptional regulator by modifying the local chromatin structure. It is localized to pericentromeric heterochromatin during interphase and mitosis (metaphase). Mutations found in ATR-X patients give rise to changes in the pattern of methylation of several highly repeated sequences, including rDNA arrays, a Y-specific satellite, and subtelomeric repeats. The severe reduction in methylation, in conjunction with the appearance of the chromatin-remodeling SNF2 helicase domain, suggests that the ATR-X protein links alterations in DNA methylation to chromatin-remodeling effects (Fig. 6.1B and D). This view is supported by observations in plants and mammals, where mutations in SWI2/SNF2-like proteins lead to a dramatic loss of genomic methylation [38, 63].

6.7
Fragile-X Syndrome

Mutation of the *FMR1* gene causes fragile-X mental retardation syndrome (MIM 309550, reviewed in [64]). The most common *FMR1* mutation is an expansion of a CCG repeat tract at the 5′ end of *FMR1* [65]. As a consequence of this expansion, the CG dinucleotides within the CCG repeat become methylated, and the associated *FMR1* gene is transcriptionally silenced (Fig. 6.2A and B). The silencing seems to be established through changes in the chromatin structure [66, 67]. The local chromatin condensation also leads to genetic stabilization of the expanded repeats.

Transfer of expanded methylated fragile-X chromosomes into mouse embryonic carcinoma cells leads to demethylation, transcriptional reactivation, and genetic instability. [68]. The methylated 5′ end of the *FMR-1* gene is associated with acetylated forms of histones H3 and H4, whereas non-expanded repeats are covered mainly by deacetylated histones. Treatment of fragile-X cells with the demethylation-inducing agent 5-aza-2′-deoxycytidine (5-aza-dC) resulted in reassociation of the acetylated histones H3 and H4 with *FMR-1* and transcriptional reactivation.

In summary, *FMR-1* silencing seems to primarily depend on DNA methylation of the expanded repeat. This is one of the best-characterized examples showing that DNA sequence alterations modify the epigenetic coding of a sequence, leading to a diseased phenotype.

6.8
Imprinting Syndromes

Imprinted genes are transcriptionally silenced by epigenetic marks (imprints) on one parental allele. Imprints are established in the gametes in a parent in an origin-specific manner, and these epigenetic differences are "heritable" throughout development of the offspring [4, 69–72]. Most of the 40 imprinted genes identified so far

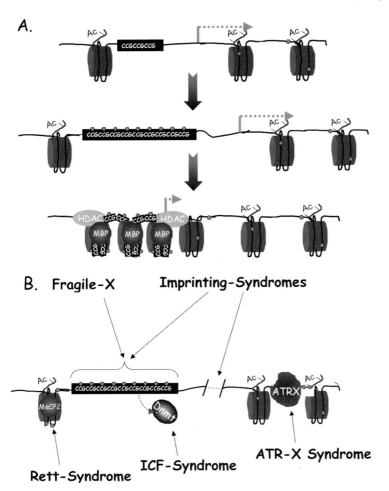

Fig. 6.2 Summary of diseases causing mutations/epimutations. **A.** A genetic change, the expansion of a CCG repeat, leads to *de novo* methylation, chromatin condensation and epigenetic silencing of the gene. **B.** Mutations/alterations found in the context of the various epigenetic diseases. Arrows indicate the connection of the mutated entity. Fragile-X and Imprinting-Syndromes shown in the top of the figure are caused by mutations or epimutations in the respective genes (deletions, expansions, chromosomal duplications, *de novo* methylation) causing changes in gene expression. The lower part of the figure highlights Syndromes (Rett-, ICF- and ATR-X Syndromes) in which the enzymes involved in epigenetic control are mutated.

are clustered in distinct genomic domains. All imprinting clusters contain CpG-rich imprinting-control elements (or imprinting centers, ICs) that are differentially methylated on the maternal and paternal chromosomes. The expression of several genes within the cluster is controlled by ICs, and targeted deletions of these lead to abnormal methylation and expression of neighboring genes.

Individual imprinting centers control the imprinting of neighboring genes in diverse ways. ICs may contain methylation-dependent binding sites for factors that regulate the domain structure. Other ICs overlap with promoters of noncoding RNA transcripts, which are usually in antisense orientation to the neighboring imprinted coding gene. The expression of most imprinted genes is tissue specific. Although the ICs are differentially methylated in all tissues, they seem to influence the imprinted expression in only some tissues and developmental stages, and no imprinting effects are observed in others. The reason for these differential epigenetic effects remains unclear. A tissue in which most imprinting effects are observed is the placenta. Normal imprinted expression of genes seems to balance the exchange of nutrients between the mother and the embryo through the placenta [73].

Imprinting plays an important role in normal development. Besides regulating aspects of growth, imprinted genes appear to also modulate behavior [72, 74]. Several developmental and behavioral disorders in humans are associated with genetic alterations in imprinted genes or their abnormal epigenetic programming [75] (see also http://cancer.otago.ac.nz:80/IGC/Web/home.html). In all these diseases, the development of symptoms depends on the parental origin of mutations or epigenetic alterations in the gene clusters. The best characterized imprinting syndromes are the Beckwith-Wiedemann syndrome (BWS) on chromosome 11p and the Prader-Willi/Angelman syndromes on chromosome 15q [4, 76]. The simplest type of genetic or epigenetic alteration in imprinting syndromes is caused by uniparental disomies (UPDs), i.e., the inheritance of two chromosomal copies of the locus from one parent. As a consequence of such UPD, the expression of imprinted genes is increased or completely absent. In BWS, a paternal UPD of chromosome 11p15.5 leads to biallelic expression of the insulin-like growth factor 2 gene (*Igf2*) and loss of expression of $p57^{KIP2}$(CDKN1C). As a consequence of such deregulated expression, BWS patients develop pre- and postnatal overgrowth syndromes, which are frequently accompanied by exomphalos and a predisposition to early childhood tumors.

Besides UPDs, genetic alterations such as translocations or deletions may also affect imprinted gene expression [77–79]. Deletions and translocations disrupt or dislocate the main regulatory elements, the ICs, causing alterations in DNA-methylation and misexpression of the neighboring imprinted genes (Fig. 6.1B).

The occurrence of epigenetic changes (imprinting defects) as a consequence of chromosomal deletions is most frequently observed in patients with Prader-Willi and Angelman syndromes. The imprinted genes responsible for these diseases are clustered on chromosome 15q11–13 in the human [4]. The Prader-Willi syndrome (PWS) is characterized by obesity, short stature, and mild mental retardation; Angelman syndrome (AS) by ataxia, hyperactivity, severe mental retardation with lack of speech, and a tendency to inappropriate laughter. The most common types of deletions in PWS and AS cover large chromosomal regions on the paternal allele (PWS patients) or on the maternal allele (AS patients). A few microdeletions have been found in some PWS and AS patients which define two elements in the IC of this cluster of deletions [80]. Parts of this IC are differentially methylated, regulating the expression of a set of neighboring genes [81]. The establishment of this imprint is rather complex and seems to occur around or shortly after the time of fertilization [82]. AS appears to be caused by either mutations in or loss of expression of the

UBE3A gene [83]. UBE3A codes for a ubiqitin protein ligase and is imprinted only in the brain. Its lack of expression in AS is caused by maternal deletions that include the 3′ part of the IC. The PWS phenotypes are caused by multiple paternally expressed imprinted genes in this region. Misexpression of these genes in PWS is caused by deletions of the paternal chromosomes that include the 3′ part of the IC. As a consequence of such deletions, the promoters of the neighboring genes become de-novo methylated and silenced.

Imprinting defects are also frequently associated with cancers, including neuroblastoma (maternal chromosome 1p36 and paternal chromosome 2), acute myoblastic leukemia (paternal chromosome 7), Wilms' tumor (maternal chromosome 11p15.5), rhabdomyosarcoma (maternal chromosome 11p15.5), sporadic osteosarcoma (maternal chromosome 13), and glomus tumors (paternal chromosomes 11q13 and 11q22.3-q23.3) (reviewed in [84]). Because some imprinted genes may act as functionally haploid tumor-suppressor genes, mutations or loss of the active copy or loss of expression of the nonimprinted allele may cause increased cancer susceptibility. Interestingly, loss of imprinting or loss of heterozygosity of the IGF2 gene is common in various tumors [85]. The imprinted IGF2R is also mutated in various cancers [86, 87].

A remarkable feature of many imprinting disorders and syndromes is the occurrence of epigenetic changes (epimutations) without any obvious accompanying genetic changes in the imprinted genes themselves. Such diseases are most likely caused by false epigenetic programming in the germ line or shortly after fertilization. Recently, two new interesting examples of such incorrect reprogramming have been described. Weksberg and colleagues [88] found loss of imprinting at imprinting centers in monozygotic twins discordant for BWS. The genetic identity of both individuals suggests that the loss of imprinting is caused by accidental false epigenetic programming during early development (see also Reik and Dean, Chapter 4). In the other report, Judson and colleagues [89] characterized epigenetic alteration in imprinted genes in a maternally inherited form of misconception (hydatidiform mole). The hydatidiform mole conceptuses have abnormal maternal imprints. These findings imply that trans-acting factors controlling the setting of imprints in the human female germ line are affected by the inherited mutation(s).

In conclusion, imprinting syndromes can be caused by a variety of genetic alterations or defective epigenetic programming effects. Both ultimately lead to abnormal expression of a number of imprinted genes and the development of complex syndromes. The results obtained from studies on imprinted genes and diseases are likely to have strong implications for an understanding of the molecular basis also of non-imprinted disorders caused by epimutations.

References

1 REIK, W.; DEAN, W.; WALTER, J., *Science* 2001, *293*, 1089–1093.

2 WOLFFE, A. P.; MATZKE, M. A., *Science* 1999, *286*, 481–486.

3 JONES, P. A.; BAYLIN, S. B., *Nature Rev. Genet.* 2002, *3*, 415–428.

4 NICHOLLS, R. D.; KNEPPER, J. L., *Annu. Rev. Genomics Hum. Genet.* 2001, *2*, 153–175.

5 MAHER, E. R.; REIK, W., *J. Clin. Invest.* 2000, *105*, 247–252.

6 BIRD, A. P.; WOLFFE, A. P., *Cell* 1999, *99*, 451–454.

7 NAKAO, M., *Gene* 2001, *278*, 25–31.

8 CHEN, R. Z.; PETTERSSON, U.; BEARD, C.; JACKSON-GRUSBY, L.; JAENISCH, R., *Nature* 1998, *395*, 89–93.

9 MATZKE, M.; MATZKE, A. J.; KOOTER, J. M., *Science* 2001, *293*, 1080–1083.

10 VAUCHERET, H.; FAGARD, M., *Trends Genet.* 2001, *17*, 29–35.

11 SLEUTELS, F.; ZWART, R.; BARLOW, D. P., *Nature* 2002, *415*, 810–813.

12 HOLLIDAY, R., *Mutat Res*, 1991, *250*, 351–363.

13 BESTOR, T. H., *Hum. Mol. Genet.* 2000, *9*, 2395–2402.

14 LI, E.; BESTOR, T. H.; JAENISCH, R., *Cell* 1992, *69*, 915–926.

15 OKANO, M.; BELL, D. W.; HABER, D. A.; LI, E., *Cell* 1999, *99*, 247–257.

16 BESTOR, T. H.; INGRAM, V. M. *Proc. Natl. Acad. Sci. USA* 1983, *80*, 5559–5563.

17 LEONHARDT, H.; PAGE, A. W.; WEIER, H. U.; BESTOR, T. H., *Cell* 1992, *71*, 865–873.

18 CHUANG, L. S.; IAN, H. I.; KOH, T. W.; NG, H. H.; XU, G.; LI, B. F., *Science* 1997, *277*, 1996–2000.

19 ROUNTREE, M. R.; BACHMAN, K. E.; BAYLIN, S. B., *Nature Genet.* 2000, *25*, 269–277.

20 ROBERTSON, K. D.; UZVOLGYI, E.; LIANG, G.; TALMADGE, C.; SUMEGI, J.; GONZALES, F. A.; JONES, P. A., *Nucleic Acids Res.* 1999, *27*, 2291–2298.

21 WALSH, C. P.; CHAILLET, J. R.; BESTOR, T. H., *Nature Genet.* 1998, *20*, 116–117.

22 RHEE, I.; BACHMAN, K. E.; PARK, B. H.; JAIR, K. W.; YEN, R. W.; SCHUEBEL, K. E.; CUI, H.; FEINBERG, A. P.; LENGAUER, C.;

KINZLER, K. W.; BAYLIN, S. B.; VOGEL-STEIN, B., *Nature* 2002, *416*, 552–556.

23 LIANG, G.; CHAN, M. F.; TOMIGA-HARA, Y.; TSAI, Y. C.; GONZALES, F. A.; LI, E.; LAIRD, P. W.; JONES, P. A., *Mol. Cell Biol.* 2002, *22*, 480–491.

24 ROBERTSON, K. D.; AIT-SI-ALI, S.; YOKO-CHI, T.; WADE, P. A.; JONES, P. L.; WOLFFE, A. P., *Nature Genet.* 2000, *25*, 338–342.

25 PRADHAN, S.; KIM, G. D., *EMBO J.* 2002, *21*, 779–788.

26 FUKS, F.; BURGERS, W. A.; GODIN, N.; KASAI, M.; KOUZARIDES, T., *EMBO J.* 2001, *20*, 2536–2544.

27 OSWALD, J.; ENGEMANN, S.; LANE, N.; MAYER, W.; OLEK, A.; FUNDELE, R.; DEAN, W.; REIK, W.; WALTER, J., *Curr. Biol.* 2000, *10*, 475–478.

28 YODER, J. A.; WALSH, C. P.; BESTOR, T. H., *Trends Genet.* 1997, *13*, 335–340.

29 BALLESTAR, E.; WOLFFE, A. P., *Eur. J. Biochem.* 2001, *268*, 1–6.

30 RICHARDS, E. J.; ELGIN, S. C., *Cell* 2002, *108*, 489–500.

31 FUKS, F.; BURGERS, W. A.; BREHM, A.; HUGHES-DAVIES, L.; KOUZARIDES, T., *Nature Genet.* 2000, *24*, 88–91.

32 NAN, X.; NG, H. H.; JOHNSON, C. A.; LAHERTY, C. D.; TURNER, B. M.; EISEN-MAN, R. N.; BIRD, A., *Nature* 1998, *393*, 386–389.

33 NG, H. H.; ZHANG, Y.; HENDRICH, B.; JOHNSON, C. A.; TURNER, B. M.; ERDJU-MENT-BROMAGE, H.; TEMPST, P.; REIN-BERG, D.; BIRD, A., *Nature Genet.* 1999, *23*, 58–61.

34 FUJITA, N.; TAKEBAYASHI, S.; OKU-MURA, K.; KUDO, S.; CHIBA, T.; SAYA, H.; NAKAO, M.. *Mol. Cell Biol.* 1999, *19*, 6415–6426.

35 WADE, P. A.; GEGONNE, A.; JONES, P. L.; BALLESTAR, E.; AUBRY, F.; WOLFFE, A. P., *Nature Genet.* 1999, *23*, 62–66.

36 NG, H. H.; JEPPESEN, P.; BIRD, A., *Mol. Cell Biol.* 2000, *20*, 1394–1406.

37 JEDDELOH, J. A.; STOKES, T. L.; RI-CHARDS, E. J., *Nature Genet.* 1999, *22*, 94–97.

38 DENNIS, K.; FAN, T.; GEIMAN, T.;

Yan, Q.; Muegge, K., *Genes Dev.* **2001**, *15*, 2940–2944.

39 Tamaru, H.; Selker, E. U., *Nature* **2001**, *414*, 277–283.

40 Keohane, A. M.; Lavender, J. S.; O'Neill, L. P.; Turner, B. M., *Dev. Genet.* **1998**, *22*, 65–73.

41 Heard, E.; Rougeulle, C.; Arnaud, D.; Avner, P.; Allis, C. D.; Spector, D. L., *Cell* **2001**, *107*, 727–738.

42 Fan, G.; Beard, C.; Chen, R. Z.; Csankovszki, G.; Sun, Y.; Siniaia, M.; Biniszkiewicz, D.; Bates, B.; Lee, P. P.; Kuhn, R.; Trumpp, A.; Poon, C.; Wilson, C. B.; Jaenisch, R., *J. Neurosci.* **2001**, *21*, 788–797.

43 Takizawa, T.; Nakashima, K.; Namihira, M.; Ochiai, W.; Uemura, A.; Yanagisawa, M.; Fujita, N.; Nakao, M.; Taga, T., *Dev. Cell* **2001**, *1*, 749–758.

44 Jones, P. L.; Veenstra, G. J.; Wade, P. A.; Vermaak, D.; Kass, S. U.; Landsberger, N.; Strouboulis, J.; Wolffe, A. P., *Nature Genet.* **1998**, *19*, 187–191.

45 Free, A.; Wakefield, R. I.; Smith, B. O.; Dryden, D. T.; Barlow, P. N.; Bird, A. P., *J. Biol. Chem.* **2001**, *276*, 3353–3360.

46 Lee, S. S.; Wan, M.; Francke, U., *Brain Dev.* **2001**, *23* Suppl 1, S138–143.

47 Amir, R. E.; Van den Veyver, I. B.; Wan, M.; Tran, C. Q.; Francke, U.; Zoghbi, H. Y., *Nature Genet.* **1999**, *23*, 185–188.

48 Nan, X.; Bird, A., *Brain Dev.* **2001**, *23* Suppl 1, S32–37.

49 Chen, R. Z.; Akbarian, S.; Tudor, M.; Jaenisch, R., *Nature Genet.* **2001**, *27*, 327–331.

50 Guy, J.; Hendrich, B.; Holmes, M.; Martin, J.E.; Bird, A., *Nature Genet.* **2001**, *27*, 322–326.

51 Huppke, P.; Bohlander, S.; Kramer, N.; Laccone, F.; Hanefeld, F., *Neuropediatrics* **2002**, *33*, 105–108.

52 Jeanpierre, M.; Turleau, C.; Aurias, A.; Prieur, M.; Ledeist, F.; Fischer, A.; Viegas-Pequignot, E., *Hum. Mol. Genet.* **1993**, *2*, 731–735.

53 Miniou, P.; Jeanpierre, M.; Blanquet, V.; Sibella, V.; Bonneau, D.; Herbelin, C.; Fischer, A.; Niveleau, A.; Viegas-Pequignot, E., *Hum. Mol. Genet.* **1994**, *3*, 2093–2102.

54 Xu, G. L.; Bestor, T. H.; Bourc'his, D.; Hsieh, C. L.; Tommerup, N.; Bugge, M.; Hulten, M.; Qu, X;, Russo, J. J.; Viegas-Pequignot, E., *Nature* **1999**, *402*, 187–191.

55 Hansen, R. S.; Wijmenga, C.; Luo, P.; Stanek, A. M.; Canfield, T. K.; Weemaes, C. M.; Gartler, S. M., *Proc. Natl. Acad. Sci. USA* **1999**, *96*, 14412–14417.

56 Ehrlich, M.; Tsien, F.; Herrera, D.; Blackman, V.; Roggenbuck, J.; Tuck-Muller, C. M., *J. Med. Genet.* **2001**, *38*, 882–884.

57 Miniou, P.; Jeanpierre, M.; Bourc'his, D.; Coutinho Barbosa, A. C.; Blanquet, V.; Viegas-Pequignot, E., *Hum. Genet.* **1997**, *99*, 738–745.

58 Ehrlich, M.; Buchanan, K. L.; Tsien, F.; Jiang, G.; Sun, B.; Uicker, W.; Weemaes, C. M.; Smeets, D.; Sperling, K.; Belohradsky, B. H.; Tommerup, N.; Misek, D. E.; Rouillard, J. M.; Kuick, R.; Hanash, S. M., *Hum. Mol. Genet.* **2001**, *10*, 2917–2931.

59 Bachman, K. E.; Rountree, M. R.; Baylin, S. B., *J. Biol. Chem.* **2001**, *276*, 32282–32287.

60 Picketts, D. J.; Higgs, D. R.; Bachoo, S.; Blake, D. J.; Quarrell, O.W.; Gibbons, R. J., *Hum. Mol. Genet.* **1996**, *5*, 1899–1907.

61 Gibbons, R. J.; Higgs, D. R., *Amer. J. Med. Genet.* **2000**, *97*, 204–212.

62 Pask, A.; Renfree, M. B.; Marshall Graves, J. A., *Proc. Natl. Acad. Sci. USA* **2000**, *97*, 13198–13202.

63 Gibbons, R. J.; McDowell, T. L.; Raman, S.; O'Rourke, D. M.; Garrick, D.; Ayyub, H.; Higgs, D. R., *Nature Genet.* **2000**, *24*, 368–371.

64 Oberle, I.; Rousseau, F.; Heitz, D.; Kretz, C.; Devys, D.; Hanauer, A.; Boue, J.; Bertheas, M. F.; Mandel, J. L., *Science* **1991**, *252*, 1097–1102.

65 Jin, P.; Warren, S. T., *Hum. Mol. Genet.* **2000**, *9*, 901–908.

66 Coffee, B.; Zhang, F.; Warren, S. T.; Reines, D., *Nature Genet.* **1999**, *22*, 98–101.

67 Chiurazzi, P.; Neri, G., *Brain Res. Bull.* **2001**, *56*, 383–387.

68 Wohrle, D.; Salat, U.; Hameister, H.;

VOGEL, W.; STEINBACH, P., Amer. J. Hum. Genet. 2001, 69, 504–515.

69 SLEUTELS, F.; BARLOW, D.P.; LYLE, R., Curr. Opin. Genet. Dev. 2000, 10, 229–233.

70 PAULSEN, M.; FERGUSON-SMITH, A. C., J. Pathol. 2001, 195, 97–110.

71 FERGUSON-SMITH, A. C.; SURANI, M. A., Science 2001, 293, 1086–1089.

72 REIK, W.; WALTER, J., Nature Reviews Genetics 2001, 2, 21–32.

73 CONSTANCIA, M.; HEMBERGER, M.; HUGHES, J.; DEAN, W.; FERGUSON-SMITH, A.; FUNDELE, R.; STEWART, F.; KELSEY, G.; FOWDEN, A.; SIBLEY, C.; REIK, W., Nature 2002, 417, 945–948.

74 KEVERNE, E. B., Prog. Brain Res. 2001, 133, 279–285.

75 MORISON, I. M.; REEVE, A. E., Molecular Medicine Today 1998, 4, 110–115.

76 REIK, W.; MAHER, E. R., Trends Genetics 1997, 13, 330–334.

77 REID, L. H.; DAVIES, C.; COOPER, P. R.; CRIDER-MILLER, S. J.; SAIT, S. N.; NOWAK, N. J.; EVANS, G.; STANBRIDGE, E. J.; DEJONG, P.; SHOWS, T. B.; WEISSMAN, B. E.; HIGGINS, M. J., Genomic, 1997, 43, 366–375.

78 LEE, M. P.; DEBAUN, M.; RANDHAWA, G.; REICHARD, B. A.; ELLEDGE, S. J.; FEINBERG, A. P.. Amer. J. Hum. Gen. 1997, 61, 304–309.

79 BROWN, K. W.; VILLAR, A. J.; BICKMORE, W.; CLAYTON, S. J.; CATCHPOOLE, D.; MAHER, E. R.; REIK, W., Hum. Mol. Gen. 1996, 5, 2027–2032.

80 BUITING, K.; SAITOH, S.; GROSS, S.; DITTRICH, B.; SCHWARTZ, S.; NICHOLLS, R. D.; HORSTHEMKE, B., Nature Genetics 1995, 9, 395–400.

81 REIS, A.; DITTRICH, B.; GREGER, V.; BUITING, K.; LALANDE, M.; GILLESSEN, K. G.; ANVRET, M.; HORSTHEMKE, B., Amer. J. Hum. Gen. 1994, 54, 741–747.

82 EL-MAARRI, O.; BUITING, K.; PEERY, E. G.; KROISEL, P. M.; BALABAN, B.; WAGNER, K.; URMAN, B.; HEYD, J.; LICH, C.; BRANNAN, C. I.; WALTER, J.; HORSTHEMKE, B., Nat Genet. 2001, 27, 341–344.

83 KISHINO, T.; LALANDE, M.; WAGSTAFF, J., Nat Genet. 1997, 15, 70–73.

84 FALLS, J. G.; PULFORD, D. J.; WYLIE, A. A.; JIRTLE, R. L., Amer. J. Pathol. 1999, 154, 635–647.

85 MOULTON, T.; CHUNG, W. Y.; YUAN, L.; HENSLE, T.; WABER, P.; NISEN, P.; TYCKO, B., Med. Pediatric Oncol. 1996, 27, 476–483.

86 KILLIAN, J. K.; OKA, Y.; JANG, H. S.; FU, X.; WATERLAND, R. A.; SOHDA, T.; SAKAGUCHI, S.; JIRTLE, R. L., Hum. Mutat. 2001, 18, 25–31.

87 OKA, Y.; WATERLAND, R. A.; KILLIAN, J. K.; NOLAN, C. M.; JANG, H. S.; TOHARA, K.; SAKAGUCHI, S.; YAO, T.; IWASHITA, A.; YATA, Y.; TAKAHARA, T.; SATO, S.; SUZUKI, K.; MASUDA, T.; JIRTLE, R. L., Hepatology 2002, 35, 1153–1163.

88 WEKSBERG, R.; SHUMAN, C.; CALUSERIU, O.; SMITH, A. C.; FEI, Y. L.; NISHIKAWA, J.; STOCKLEY, T. L.; BEST, L.; CHITAYAT, D.; OLNEY, A.; IVES, E.; SCHNEIDER, A.; BESTOR, T. H.; LI, M.; SADOWSKI, P.; SQUIRE, J. Hum. Mol. Genet. 2002, 11, 1317–1325.

89 JUDSON, H.; HAYWARD, B. E.; SHERIDAN, E.; BONTHRON, D. T., Nature 2002, 416, 539–542.

7
Liver or Broccoli? Food's Lasting Effect on Genome Methylation

Me

Folate

Mutant
unstable DNA

7

Liver or Broccoli? Food's Lasting Effect on Genome Methylation*

MICHAEL FENECH

Summary

Damage to the genetic material (DNA) in cells and inappropriate expression of genes increase the risk of developmental defects, cancer, and accelerated aging. Over the past four decades, compelling evidence has emerged that the essential micronutrients (or vitamins) folate, vitamin B12, choline, and methionine play a central role in maintaining the stability of DNA by providing and enabling donation of carbon atoms (or methyl groups) for synthesis of the DNA bases (which determine the genetic code and are required for DNA repair and replication) and for the maintenance of methylation patterns on DNA (which determine which genes are expressed). Current knowledge suggests that the levels of folate and vitamin B12 required to minimize the rate of DNA damage in human populations is probably well above the average dietary intake that was previously considered adequate for preventing vitamin-deficiency diseases such as anemia. These results suggest that optimizing genome methylation by dietary means or by supplementation should prove to be an effective strategy for preventing diseases caused by genomic instability. Important dietary sources of folate, vitamin B12, choline, and methionine are reviewed.

7.1

Introduction

The vitamins folate acid and vitamin B12, as well as methionine and choline, play an important role in maintaining the stability of DNA structure and preventing harmful alterations to the genetic code. They do this by providing single carbon atoms (methyl groups) for the synthesis of bases in DNA (e.g., thymine from uracil) or to

---

* Substantial parts of this paper, including Fig. 7.1 and Table 7.2, were updated and/or reprinted from Mutation Research, vol. 475, Fenech M., "The role of folic acid and vitamin B12 in genomic stability of human cells", pages 57–67, 2001, with permission of Elsevier Science. – Abbreviations: HC = homocysteine, MN = micro- nucleus, MNed = micronucleated, RDA = recommended dietary allowance, SSB = single-strand break, DSB = double-strand break, RBCs = red blood cells. – Key to unit conversion: 1 ng/ml folic acid = 2.26 nmol/L folic acid; 1 pg/ml vitamin B12 = 0.74 pmol/L vitamin B12.

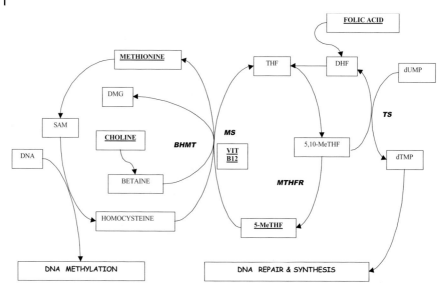

Fig. 7.1 The main metabolic pathways by which folate, vitamin B12, choline and methionine affect DNA methylation, synthesis, and repair. BHMT, betaine:homocysteine methyltransferase; DHF, dihdrofolate; DMG, dimethylglycine; 5-MeTHF, 5-methyltetrahydrofolate; 5,10-MeTHF, 5,10-methylene-tetrahydrofolate; MS, methionine synthase; MTHFR, methylenetetrahydrofolate reductase; SAM, S-adenosyl methionine; THF, tetrahydrofolate; TS, thymidylate synthase; VIT B12, vitamin B12

keep one of these bases, cytosine, in its methylated state to enable appropriate control of gene expression [1] (Fig. 7.1).

Under conditions of folate deficiency, uracil accumulates and, as a result, is incorporated into DNA instead of thymine [2]. Good evidence suggests that excessive incorporation of uracil into DNA not only leads to alteration of the genetic code but may also result in breakage and rearrangement of DNA, which can be observed under the microscope as broken chromosomes or small nuclear fragments called micronuclei [3, 4]. The mutagenic effects of uracil are underscored by the observation that, of eight known human enzymes required for removing abnormal DNA bases, four (UNG, TDG, hSMUG1, MBD4) are dedicated to the removal of uracil [5]. Folate and vitamin B12 are also required for the synthesis of methionine and S-adenosyl methionine (SAM). SAM is the common methyl donor required for maintenance of methylation patterns in DNA that determine gene expression and DNA structure [6]. When the concentration of vitamin B12 is low, folate becomes unavailable for the synthesis of thymine from uracil or for the synthesis of methionine from homocysteine. Therefore, deficiencies in folate and vitamin B12 can lead to (i) elevated DNA damage rate and altered methylation of DNA, both of which are important risk factors for cancer , (ii) an increased level of homocysteine, an important risk factor for cardiovascular disease, and (iii) reduced cell division, which can lead to anemia [3–7]. These same defects may also play an important role in developmental and neurological abnormalities [3, 4].

The blood levels of folate and vitamin B12 required to prevent anemia and high homocysteine concentrations are defined; however, it is still uncertain whether such accepted levels of sufficiency are in fact adequate to also minimize chromosome damage rates and optimize DNA methylation status. In this review, evidence is provided from in vitro studies with human cells and in vivo cross-sectional and intervention studies in humans to identify the concentration or intake level at which potential genotoxic effects of low folate and vitamin B12 status may be prevented. In addition, the potential impact of genetic polymorphisms in key transport molecules and enzymes required for the metabolism of folic acid and vitamin B12 are discussed as factors that should be considered when determining recommended dietary allowances (RDA) for the intake of these vitamins based on genomic stability (i.e., prevention of DNA damage and abnormal DNA methylation levels) [8].

Most studies relating to genome methylation and diet have focused on folate and vitamin B12. However, before proceeding to describe the data for folate and vitamin B12, it is also important to consider the other important dietary sources of methyl groups, namely, methionine and choline. Methionine is the immediate precursor of the methyl donor SAM, required for maintenance methylation of DNA. The effect of concentration and/or intake of methionine and its metabolites with respect to genome stability, DNA methylation, cell division, and cancer control is emerging as an important research area [9–13]. However, there is insufficient direct evidence in humans to enable an in-depth discussion and recommendation now. The precise contribution of choline to DNA methylation in humans is unknown; however, inadequate choline in the diet, on its own, induces a deficit in DNA methylation and increased risk of hepatocarcinomas in rodents [14–20]. This is not surprising, given that each choline molecule is a source of three methyl groups, which become available when choline is oxidized (mainly in the liver and kidney) to betaine. Betaine can donate its methyl groups for the conversion of homocysteine to methionine, which serves as an alternative to methionine synthesis involving 5-methyltetrahdrofolate and vitamin B12 (Fig. 7.1). When choline is deficient, SAM is consumed in the de-novo biosynthesis of choline by methylation of phosphotidylethanolamine, which may increase the demand for folate and methionine [14–17].

7.2
Important Dietary Sources of Folate, Vitamin B12, Choline, and Methionine

It is becoming increasingly evident that large proportions (10%–70%) of populations in developed countries are not meeting the current recommended dietary allowances (RDA) of folate and vitamin B12 [21–24]. The extent of choline and methionine deficiencies is unknown, because RDAs for these nutrients have not been set. These dietary deficiencies could be due to insufficient knowledge of the micronutrient content of foods and/or poor choice of foods. It is therefore useful to consider important dietary sources of folate, vitamin B12, choline, and methionine [16, 25–27]. This knowledge is essential for anyone who intends to achieve the RDA of folate ($400 \ \mu g \ d^{-1}$) and vitamin B12 ($2.4 \ \mu g \ d^{-1}$) and to have an adequate intake of choline

Tab. 7.1. Examples of dietary sources of folate, vitamin B12, methionine, and choline (data from [16, 25–27])

Folate – µg/100g	Vitamin B12 – µg/100g[a]	Methionine – mg/100g	Choline – mg/100g[b]
Meats	*Meats*	*Meats*	*Meats*
• Chicken liver (fried) – 1385	• Ox liver (stewed) – 110	• Beef steak (fried) – 840	• Beef liver (cooked) – 711
• Beef liver (fried) – 1057	• Lamb liver (fried) – 81	• Beef (boiled) – 750	• Beef steak (cooked) – 92
• Pig liver (fried) – 540	• Chicken liver (fried) – 49	• Chicken liver (fried) – 500	*Vegetables*
• Ground beef (stewed) – 16	• Lamb kidney (fried) – 79	• Pig liver (stewed) – 610	• Cauliflower (cooked) – 59
• Beef steak (fried) – 15	• Beef steak (fried) – 2	*Fish*	• Potato (cooked) – 12
Fish	• Ground beef (fried) – 2	• Cod (baked) – 620	• Tomato – 7
• Salmon (steamed) – 29	*Fish*	• Haddock (fried) – 660	• Cucumber – 5
• Sardines (canned) – 26	• Salmon (steamed) – 5	*Cheese*	• Iceberg lettuce – 43.2
Cereals	• Tuna (canned) – 5	• Parmesan – 990	*Fruits*
• Aleurone flour – 500	• Sardines (canned) – 28	• Cheddar – 730	• Grape juice (canned) – 7
• Wheat germ – 90	*Other*	*Nuts*	• Apple – 4
• Wheat bran – 47	• Condensed milk – 1	• Brazil nuts – 800	• Banana – 4
• Wholemeal bread – 23		• Peanuts – 320	• Orange – 10
Fruits		*Fruits*	*Other*
• Strawberry – 65		• Avocados – 67	• Whole-wheat bread – 18
• Kiwifruit – 23		• Currants (dried) – 57	• Milk (cow) – 5
• Orange – 18		*Vegetables*	• Egg (cooked) – 754
Vegetables		• Cauliflower (boiled) – 31	• Peanuts – 133
• Brussel sprouts (cooked) – 87		• Lentils (boiled) – 61	
• Spinach (cooked) – 83		• Mushrooms (fried) – 81	
• Broccoli (cooked) – 65		• Peas (boiled) – 48	
• Lettuce (fresh) – 58			

a Plant foods such as vegetables, fruit, cereals, and nuts do not contain vitamin B12.

b The figures shown for total choline represent choline content plus choline equivalents from phosphatidylcholine and sphingomyelin.

and methionine via foods rather than supplements. As indicated above, the optimal intake of methionine and choline for preventing DNA hypomethylation in humans remains unknown; however, because of the potential importance of these essential nutrients in cancer prevention and control, it may be useful to be aware of choline- and methionine-rich foods. Recent data suggest that dietary choline intake should exceed 250 mg d^{-1} to maintain adequate plasma concentrations of choline and phosphotidylcholine when folate intake is low [15], and recommended adequate intake levels have been set as 400–550 mg d^{-1} for adults, depending on gender, pregnancy, and lactation [16]. Table 7.1 lists some examples of important and less important dietary sources of folate, vitamin B12, methionine, and choline.

Liver is one of the richest sources of folate (Table 7.1). Although broccoli is among the vegetables with the highest folate content, one would have to consume approximately 600 g of cooked broccoli but only 30 g of fried chicken liver to achieve the RDA for folate. Aleurone flour, made from the aleurone layer of wheat grain, is one of the richest plant sources of folate, with the additional benefit of providing high bioavailability in humans [26]. Liver is also an excellent source of vitamin B12–10–30 g provides above the RDA level of this vitamin. In contrast, plant foods are devoid of this critical vitamin. Meat, liver, fish, cheese, and nuts have the highest contents of methionine, exceeding the concentration in fruits and vegetables by 5–10 fold. Eggs and liver are the best known sources of choline – 50–100 g of these foods is sufficient to meet the recommended adequate intake of choline. Therefore, making careful food choices can have an important impact on an individual's success in achieving adequate intake of key micronutrients required for maintenance methylation of DNA.

7.3
Evidence from in Vitro Cultures for the Role of Folic Acid in Genomic Stability of Human Cells

Fragile sites in chromosomes are expressed when human lymphocytes are cultured in the absence of folic acid and thymidine [28, 29]. Furthermore, under these conditions chromosome breakage and micronucleus (MN) expression are increased simultaneously, suggesting that a similar mechanism underlies the expression of fragile sites and chromosome breakage [28–30]. Reidy's experiments [31, 32] showed that lymphocytes cultured in folic acid-deficient medium exhibit increased levels of excision repair during G$_2$, because the cytosine arabinoside-induced chromosome aberration level (which indicates excision repair activity) was more than doubled by this treatment and further enhanced by addition of deoxyuridine. Treatment of human lymphoid cells in culture with methotrexate results in a large increase in the dUTP/dTTP ratio and a much increased incorporation rate of uracil into DNA [33]. The connection between uracil incorporation and the generation of DNA strand breaks was confirmed in more recent studies using the single-cell gel electrophoresis method; the extent of uracil incorporation was measured by treating the nuclei with uracil DNA glycosylase followed by measurement of the resulting DNA breaks [34, 35].

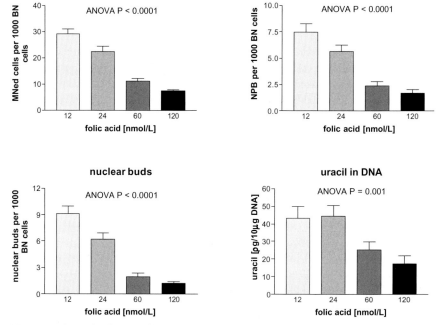

Fig. 7.2 Relationship between folic acid concentration and four different biomarkers of DNA damage. MNed = micronucleated cells, a biomarker for chromosome breakage and loss; NPB = nucleoplasmic bridges, a biomarker of chromosome rearrangement; nuclear buds = a biomarker for gene amplification. Results obtained from long-term primary lymphocyte cultures of 20 different subjects (i.e., N = 20); data from Crott et al. [37]

To provide a more in-depth understanding of the impact of folic acid deficiency on chromosomal stability, we cultured lymphocytes for 9 days in culture medium with decreasing folic acid concentrations within the physiological range (i.e., 10–120 nmol L^{-1}). We used a new multiple end-point cytokinesis-block micronucleus assay to measure micronuclei (a marker of chromosome breakage and chromosome loss), nucleoplasmic bridges (a marker of chromosome rearrangement), and nuclear buds (a marker of gene amplification), necrosis, apoptosis, nuclear division index, as well as uracil in DNA. The results of this study showed that chromosome damage rates (Fig. 7.2), uracil in DNA, and necrosis were minimized at a folic acid concentration of 120 nmol L^{-1}, and the nuclear division index was maximized at this concentration; however, no dose-related trends were found for apoptosis [36, 37].

It appears that no conclusive studies have yet been published showing a link between vitamin B12 deficiency in vitro and increased genomic instability in human cells.

7.4

Evidence from in Vivo Studies for the Role of Folate and Vitamin B12 in Genomic Stability of Human Cells

Evidence of chromosome damage in human cells in vivo from folate and vitamin B12 deficiency was first obtained from studies linking the expression of Howell-Jolly bodies in erythrocytes with megaloblastic anemias [38–40]. Howell-Jolly bodies are whole chromosomes or chromosome fragments that lag behind at anaphase during production and maturation of the red blood cell, and in fact, they are the same as micronuclei, the current term for this chromosome damage biomarker. Micronucleated erythrocytes in humans are most readily observed in splenectomised subjects, because the spleen actively filters micronucleated erythrocytes from the blood [41, 42].

A study of a 30-year-old man with Crohn's disease and a very high level of micronuclei in erythrocytes (67/1000 cells) showed that this was associated with low serum folate (1.9 ng mL^{-1}; normal range >2.5 ng mL^{-1}) and low red cell folate (70 ng mL^{-1}; normal range >225 ng mL^{-1}). Micronucleus frequency was reduced to 12/1000 cells, serum folate increased to >20 ng mL^{-1}, red cell folate increased to 1089 ng mL^{-1} after 25 days with a daily oral dose of 25 mg folinic acid [41]. One of the main observations of this study was that spontaneous MN frequencies were minimal only when serum folate levels exceeded 15–20 ng mL^{-1}, which is higher than the values accepted as normal by clinicians (i.e., 6–15 ng mL^{-1}).

A cross-sectional study of smokers (N = 30) and nonsmokers (N = 30) showed a significant inverse relationship between chromosome aberrations and blood folate status and that smoking and blood folate status are interrelated in their association with chromosome fragility. In this study of ex vivo expressed DNA damage, the cells were cultured in low-folate medium, so the results may therefore reflect the expression of fragile sites within chromosomes [43].

Another small (N = 22) cross-sectional study of the influence of blood micronutrients on micronucleus frequency in erythrocytes of splenectomised subjects selected from a larger population because they had among the highest or lowest micronucleus frequencies (N = 122) showed that the elevated micronucleus index was strongly associated only with low levels of serum folate (<4 ng mL^{-1}) and low levels of plasma B12 (<200 pg mL^{-1}). Vitamins C and E and beta-carotene did not show a strong inverse correlation with the micronucleus index [44].

Blood samples from the same splenectomised individuals [41, 44] were also analyzed for uracil content [3, 4]. The results showed that the uracil level in DNA was 70-fold higher in individuals with serum folate <4 ng mL^{-1} than in individuals with serum folate >4 ng mL^{-1}. Uracil levels in DNA were rapidly reduced (within 3 days) after daily supplementation with 5 mg folic acid in both the deficient (<4 ng mL^{-1} serum folate) and nondeficient (>4 ng mL^{-1} serum folate) groups. These changes were accompanied by corresponding reductions in erythrocyte micronucleus frequency, but over a longer time-frame [3, 4].

A folic acid depletion/repletion study (baseline: 195 µg d^{-1}; depletion: 5 weeks at 65 µg d^{-1}; repletion: 4 weeks at 111 µg d^{-1} followed by 20 d at >280 µg d^{-1}) of nine postmenopausal women in a metabolic unit showed a significant increase in micro-

nucleus frequency in lymphocytes following depletion and a decrease following repletion; the micronucleus frequency in buccal cells decreased after the repletion phase
[45]. The depletion phase in this study also resulted in increased DNA hypomethylation, increased dUTP/dTTP ratio, and decreased NAD levels in lymphocytes [46].

A cross-sectional study on buccal mucosal folate and vitamin B12 and its relation to
micronucleus frequency in buccal cells revealed that buccal mucosal folate and vitamin B12 are significantly lower in current smokers than in former smokers [47].
Although current smokers in this study were three times more likely to have micronucleated buccal cells, this chromosome damage index was not associated with localized
folate or vitamin B12 deficiency. However, elevated salivary vitamin B12 was associated with a reduced micronucleus frequency in the buccal cells. This was the first investigation of the hypothesis that epithelial cancers such as those of the cervix, lung,
bladder, and oropharyngeal region could be due to localized deficiencies in folic acid
and vitamin B12, which was suggested from the observation that megaloblastic
changes in such tissues can be corrected by folate/vitamin B12 supplementation [48].

A small number of case studies link vitamin B12 deficiency with increased levels
of chromosome aberrations [49, 50]. Of 10 patients with pernicious anemia (which is
a manifestation of vitamin B12 deficiency), 3 had elevated chromosome aberrations,
and 8 had increased levels of micronucleus frequency in bone marrow preparations
[51]. A female infant with transcobalamin II (the transporter for vitamin B12 in
plasma) deficiency showed elevated levels of aneuploidy (hypodiploidy in approximately 30% of cells) and increased chromosomal breakage in the bone marrow. The
hypodiploidy decreased to 10% of cells after 5 months of treatment with folate and
vitamin B12 supplements [52].

Combined deficiency in folic acid and vitamin B12 (i) was associated with loss of
the q arm of chromosome 7 [53] and (ii) in a series of patients produced a persistent
abnormal deoxyuridine suppression test result (which is indicative of inadequate capacity to generate dTMP) and increased frequency of chromosomes showing despiralization and chromosomal breaks [54]. The latter studies showed that it took up to
84 days after supplementation with folic acid and vitamin B12 before the deoxyuridine suppression and chromosomal morphology tests returned to normal.

With regard to the question of chromosome despiralization, it may be important
to note that the DNA methylation inhibitor, 5-azacytidine, induces distinct undercon-
densation of the heterochromatin regions of chromosomes 1, 9, 15, 16, and Y and
the specific loss of these chromosomes as micronuclei in human lymphocytes in vitro [55]. Similarly, the ICF immunodeficiency syndrome, which is caused by mutation of the DNA methyl transferase gene, is characterized by despiralization of heterochromatin of chromosomes 1, 9, and 16 and loss of this chromatin into micronuclei and nuclear blebs [56].

A cross-sectional study in Japan involving 18 college students aged 19–23 y and 15
laboratory workers aged 24–69 y, which was performed to investigate the relation of
age-adjusted micronucleus index in cytokinesis-blocked lymphocytes to serum vitamins, found a protective effect of increased folic acid that was marginally significant
(multiple regression beta value -4.00, $P = 0.06$), but no apparent protective effect associated with elevated vitamin B12 concentration. None of the subjects were vitamin

B12-deficient (<200 pg mL^{-1}) or folic acid-deficient (<3.5 ng mL^{-1}); the mean values were 544 pg mL^{-1} B12 and 7 ng mL^{-1} folate [57]. In contrast, low but not deficient levels of plasma B12 were associated with increased micronucleus frequency ($r^2 = 0.11$, P = 0.06) in nascent human erythrocytes of healthy blood donors [58].

We performed a series of investigations of the interrelationship between DNA damage in somatic cells and blood status for folate, vitamin B12, and homocysteine. We used the cytokinesis-block micronucleus method in lymphocytes, which is a reliable and sensitive biomarker of chromosome breakage and of chromosome loss that occurs spontaneously [59] or as a result of elevated exposure to genotoxins [60].

Our preliminary studies comparing DNA damage rate and micronutrient status in vegetarians and nonvegetarians had indicated a significant negative correlation between the micronucleus frequency in lymphocytes and plasma vitamin B12 status in young men [61]. Therefore, we investigated the prevalence of folate deficiency, vitamin B12 deficiency, and hyperhomocysteinemia in 64 healthy men aged 50–70 y and determined the relationship of these micronutrients to the micronucleus frequency in cytokinesis-blocked lymphocytes [62]. The serum folate concentration in 23% of the men was less than 6.8 nmol L^{-1}, 16% had red blood cell folate concentrations less than 317 nmol L^{-1}, 4.7% were vitamin B12-deficient (<150 pmol L^{-1}), and 37% had plasma homocysteine levels greater than 10 _mol L^{-1}. In all, 56% of the apparently healthy men had non-optimal values for folate, vitamin B12, or homocysteine. The micronucleus index of these men (19.2 ± 1.1, N = 34) was significantly elevated (P = 0.02) compared to that of men who had higher concentrations of folate and vitamin B12 and lower plasma homocysteine (16.3 ± 1.3, N = 30). Interestingly, the micronucleus index in men with normal concentrations of folate and vitamin B12 but homocysteine levels greater than 10 µmol L^{-1} (19.4 ± 1.7, N = 15) was also significantly higher (P = 0.05) than in men with normal folate and vitamin B12 and homocysteine less than 10 µmol L^{-1}. Micronucleus index and plasma homocysteine were also significantly (P = 0.0086) and positively correlated (r = 0.415) in those subjects who were not deficient in folate or vitamin B12. The micronucleus index was not significantly correlated with folate indices, but there was a significant (P = 0.013) negative correlation with serum vitamin B12 (r = -0.315). It was apparent that elevated homocysteine status, in the absence of vitamin deficiency, and low, but not deficient, vitamin B12 status are important risk factors for increased chromosome damage in lymphocytes.

Subsequently, we performed a cross-sectional study (N = 49 males, 57 females) and a randomized double-blind placebo-controlled dietary intervention study (N = 31, 32 per group) to determine the effect of folate and vitamin B12 (B12) on DNA damage (micronucleus formation and DNA methylation) and plasma homocysteine (HC) in young Australian adults aged 18–32 y [63]. None of the volunteers were folate deficient (i.e., red cell folate <136 nmol L^{-1}) and only 4.4% (all females) were vitamin B12-deficient (i.e., serum B12 <150 pmol L^{-1}). The cross-sectional study showed that (i) the frequency of micronucleated (MNed) cells was positively correlated with plasma HC in males (r = 0.293, P < 0.05), and (ii) in females the MNed cell frequency was negatively correlated with serum B12 (r = −0.359, P < 0.01), but (iii) there was no significant correlation between micronucleus index and red cell folate status. The results also showed that the level of unmethylated CpG (DNA) (measured using the Sss1 methylase method) was not significantly related to vitamin B12 or folate status. The dietary intervention involved supplementation with 700 µg folic acid and 7 µg vitamin B12 in wheat bran cereal for 3 months followed by 2000 µg folic acid and 20 µg vitamin B12 in tablets for a further 3 months. In the supplemented group, MNed cell frequency was significantly reduced (by 25.4%) in those subjects with initial MNed cell frequency in the high 50th percentile, but there was no change in those subjects in the low 50th percentile for initial MNed cell frequency. The reduction in MNed cell frequency was significantly correlated with serum B12 ($r = -0.49$, P < 0.0005) and plasma HC ($r = 0.39$, P < 0.006), but was not significantly related to red cell folate. DNA methylation status was not altered in the supplemented group. The greatest decrease in plasma HC (by 37%) was observed in those subjects in the supplemented group with initial plasma HC in the high 50th percentile and correlated significantly with increases in red cell folate ($r = -0.64$, P < 0.0001) but not with serum B12. The results from this study suggest that (i) MNed cell frequency is minimized when plasma HC is below 7.5 µmol L^{-1} and serum B12 is above 300 pmol L^{-1}, and (ii) dietary supplement intake of 700 µg folic acid and 7 µg vitamin B12 is sufficient to minimize MNed cell frequency and plasma homocysteine in young adults. Thus, it appears that elevated plasma HC, a risk factor for cardiovascular disease, may also be a risk factor for chromosome damage.

Other studies have shown that global DNA methylation in lymphocytes or colonic tissue is influenced by the extent of folate intake. The depletion-repletion study performed by Jacob et al. [46] with postmenopausal women in a metabolic unit showed a more than 100% increase in DNA hypomethylation after 9 weeks on low folate (56–111 µg d^{-1}) and a subsequent increase in DNA methylation after a further 3 weeks on a high-folate diet (286–516 µg d^{-1}). Fowler et al. [64] and Cravo et al. [65] showed, using the Sss1 methylase assay, that cervical and gut epithelium DNA methylation is significantly correlated with serum and tissue folate concentrations. Furthermore, Cravo et al. [65] showed that intrinsic methylation of DNA was lower in the normal colorectal mucosa of adenoma and carcinoma patients; however, supplementation with 10 mg folic acid d^{-1} for 6 months increased methylation 15-fold (P < 0.0002), and three months after cessation of therapy, methylation decreased 4-fold.

Studies on breast cancer patients at the time of disease presentation and before chemotherapy showed that elevated mutant frequencies in the *HPRT* gene occurred in those individuals with serum folate in the deficient range and that serum vitamin B12 levels correlated negatively with sister chromatid exchange levels [66]. The extent of increase in *HPRT* mutant frequency in lymphocytes of breast cancer patients after chemotherapy tended to correlate negatively with serum folate level [67]. It may also be important to note that murine cells deficient in DNA methyl transferase exhibit elevated mutation rates, due mainly to gene deletions caused by mitotic recombination or chromosome loss [68] because these events are involved in the carcinogenic process.

Another important possibility for preventing genomic instability could be the prevention of integration of oncogenic virus DNA. Prevention of hypomethylation may enable better surveillance of foreign DNA integration into human DNA, because DNA methylation appears to have evolved partly for this purpose [69]. It is interesting to note that HPV virus tends to integrate into fragile sites that may be folate dependent [70], which suggests that viral integration into DNA in vivo may be facilitated when folate status is low enough to cause fragile site expression. It is also important to note that transcription of retroviral or parasitic DNA sequences integrated into mammalian DNA is inhibited by cytosine methylation, and conversely, that demethylation may activate transcription of endogenous retroviruses; the significance of these observations is underscored by the fact that most 5-methylcytosine in the genome actually lies within parasitic, retroviral, or transposon DNA [71, 72]. Whether folate deficiency can activate transcription of retroviral DNA remains untested. Vitamin B12 may also play a direct role in preventing integration of onco-genic viruses, because cobalamin inhibits HIV integrase and the integration of HIV DNA into nuclear DNA [73]. On the basis of these results, combination treatment with folic acid and vitamin B12 supplements has been used in AIDS patients with apparent success [74].

7.5
Environmental and Genetic Factors that Determine the Bioavailability of Folate and Vitamin B12

Alcoholism is associated with significantly reduced levels of tissue folate, vitamin B12, and vitamin B6 in humans; intake of alcohol intakes of greater than $3.0 \text{g kg}^{-1} \text{d}^{-1}$ led to a doubling in the level of DNA hypomethylation in lymphocytes [75]. The reduced folate level in alcoholics may be due to reduced absorption or suboptimal dietary intake. However, if results in the rat reflect the situation in humans, then there is a good probability that microbial metabolism of alcohol can result in exceedingly high levels of acetaldehyde, which destroys folate in the intestine – this has been shown to be associated with localized folate deficiency in the colonic mucosa [76].

Reduced absorption of protein-bound B12 due to atrophic gastritis caused by auto-antibodies to gastric pareital cells and reduced absorption due to autoantibodies to intrinsic factor are two of the main causes of vitamin B12 deficiency, which may affect 10%–40% of the elderly (> 60 y) [77]. An increasingly important cause of cobalamin deficiency is exposure to nitrous oxide due to abuse [78], exposure during anesthesia [79], or occupational exposure of hospital personnel during surgery [80]. Nitrous oxide inactivates cobalamin, rendering exposed individuals effectively B12 deficient. It is interesting that hospital personnel exposed to nitrous oxide had four times the MN frequency of matched controls [80], which could be explained in part by the genotoxic effect of functional B12 deficiency.

The conversion of dietary folate and vitamin B12 to intracellular active coenzyme requires many physiological and biochemical processes, including release of bound vitamin in the stomach, intestinal uptake, blood transport by proteins, cell uptake by receptors, and enzymatic conversion to the active coenzyme [81]. For vitamin B12, at least 5 different peptides (R binder, intrinsic factor, ileal receptors, transcobalamin I, transcobalamin II) are required to deliver vitamin B12 from the gut to the tissues, and a further 4 enzymes (cblF, cblC/D, microsomal reductase, cblE/G) are necessary to convert vitamin B12 to the appropriate reduced state for function as a coenzyme with methionine synthase. For folate, a conjugase enzyme is required to deconjugate polyglutamated folate in the small intestine; and receptors are required for active uptake into the intestinal brushborder epithelium. Then it is carried by the hepatoportal circulation to the liver, where monoglutamated folate (i.e., folic acid) is reduced and methylated to form 5-methyltetrahydrofolate, which is exported to the blood. It is then taken up into cells by a receptor/pinocytosis mechanism and converted to the polyglutamated form by the activity of folyl-gamma-polyglutamate synthetase. The capacity of 5-methyltetrahydrofolate to donate its methyl group for the regeneration of methionine from homocysteine depends on the activity of methionine synthase. The activities of thymidylate synthase and methylenetetrahydrofolate reductase determine the probability of 5,10-methylenetetrahydrofolate donating its methyl group for the conversion of dUMP to dTMP.

All the above indicate that genetic defects in one or more of the key enzymes or uptake proteins can limit the bioavailability of folate and vitamin B12. It may therefore be necessary for intake of these vitamins above the RDA, to overcome uptake-related defects or reduced enzymatic activities. This need has in fact been shown in subjects defective in intrinsic factor or cobalamin reductase (vitamin B12) and in subjects defective in MTHFR (folate) [81–83]. It is also interesting that for MTHFR, polymorphisms that reduce its activity, such as the C677T mutation, may on the one hand protect against cancer and on the other hand increase the risk for developmental defects such as Down syndrome and neural tube defects [84–89]. The most plausible explanation is that the MTHFR mutation minimizes incorporation of uracil into DNA and therefore chromosome breakage and rearrangement while having only a relatively minor impact on DNA methylation – this implies that chromosome breakage and rearrangement may be more critical than hypomethylation for carcinogenesis, although this emphasis may change depending on folate intake and the extent to which hypomethylation of DNA causes aneuploidy, a potential carcinogenic event [84, 85, 90].

However, the impact of the MTHFR mutation on DNA methylation status may be important during the finely tuned development process, when the concerted and timely expression of genes is critical and possibly more sensitive to appropriate DNA methylation status. Current in vivo evidence suggests that MTHFR C677T homozygotes have lower global DNA methylation than wildtypes in lymphocytes [91]. In vitro studies suggest that, under conditions of supraphysiological concentrations of riboflavin and methionine, C677T homozygotes are equally as susceptible as wildtypes to uracil incorporation into DNA and chromosome damage induced by folate deficiency [48, 49]. Either way, the observations on C677T and A1298C MTHFR polymorphism with respect to cancer risk and developmental defects [87, 88] suggest the potential importance of folate supplements for overcoming metabolic limitations. The same should apply to vitamin B12 for defects in other key enzymes, such as methionine synthase and methionine synthase reductase [92, 93].

7.6
Recommended Dietary Allowances (RDAs) for Folate and Vitamin B12 Based on Genomic Stability

There is now increasing interest in redefining the recommended dietary allowances of minerals and vitamins, not only to prevent diseases of extreme deficiency, but also to prevent developmental abnormalities and degenerative diseases of old age and to optimize cognition [93]. Prevention of chromosome breakage and aneuploidy is an important parameter for the definition of new RDAs for micronutrients [8, 94] such as folic acid and vitamin B12, because increased rates of DNA damage are associated with increased cancer risk [95–97] and accelerated aging [98].

Table 7.2 summarizes the information from in vitro and in vivo controlled experiments in human cells and human subjects with a view to defining, based on current knowledge, the optimal concentration and dietary intake of folic acid for minimizing genomic instability. The results from a variety of DNA damage biomarkers suggest that levels of folic acid intake above the current RDA are required to minimize DNA damage. Furthermore, the current sufficiency levels of folate in plasma (i.e., 2.2 ng mL^{-1} or 4.9 nmol L^{-1}) and red blood cells (i.e., 132 ng mL^{-1} or 298 nmol L^{-1}), which are based on prevention of anemia [99], are much lower than the average concentration levels at which DNA damage is minimized, i.e., 21 ng mL^{-1} (range 7.3–53.0 ng mL^{-1}) for plasma and 464 ng mL^{-1} (range 313–600 ng mL^{-1}) for red blood cells (Table 7.2). In this regard it is interesting that the red cell folate concentration that corresponds to minimization of risk of neural tube defects in the unborn child is approximately 400 ng mL^{-1} [100, 101], which is similar in magnitude to that which appears to be required for minimizing genomic instability. The data from our intervention studies [62, 63] also suggest that the current sufficiency level of vitamin B12

Tab. 7.2 Concentration and dietary intake of folic acid that minimises genomic instability in human tissue[a]

Genomic instability biomarker	Concentration in culture medium – in vitro (ng mL^{-1})	Concentration in plasma – in vivo (ng mL^{-1})	Concentration in RBCs – in vivo (ng mL^{-1})	Daily dietary supplement intake ($\mu g\ d^{-1}$)
SSB/DSB – Comet Assay	100 [35]			
Micronuclei	80 [29][b]	15 [41]	600 [41]	5000 [41]
	53 [37]	7 [45]		228 [45]
			313 [63]	700 [63][c]
Nucleoplasmic Bridges	53 [37]			
Nuclear Buds	53 [37]			
Uracil in DNA	53 [36]	53 [3,4]	480 [3,4]	5000 [3,4]
CpG		24 [65]		10000 [65]
Hypomethylation		7 [46]		516 [46]

[a] 1 ng mL^{-1} of folic acid = 2.26 nmol L^{-1}.
[b] In the presence of thymidine (4.0 mg L^{-1}).
[c] Together with 7 μg d^{-1} vitamin B12.
Numbers in brackets are references.

in plasma for prevention of anemia (i.e., 150 pmol L^{-1} or 203 pg mL^{-1}) is almost half the level at which the micronucleus index in lymphocytes is minimized (i.e., 300 pmol L^{-1} or 406 pg mL^{-1}). Intakes of 3.5 times the current Australian RDA were required to achieve this level in plasma.

Only with careful choice and sufficient intake of folate-rich foods, such as aleurone flour, certain fruits and vegetables, and of vitamin B12-rich foods, such as liver, is it possible to achieve the required above-RDA intake of folate and vitamin B12 [101, 102]. The use of fortified foods or tablet supplements may be more practical for achieving these levels of intake [103]. Combining folic acid with vitamin B12 in supplements or in a fortification program not only has the benefit of maximizing the impact of these vitamins, because vitamin B12 makes folate more bioavailable for synthesis of both dTMP and methionine, but it also enhances the homocysteine-lowering capacity of folate [63] and prevents the possibility of masking neuropathies (associated with pernicious anemia) when folate-only supplements are taken by individuals who have an underlying vitamin B12 deficiency [104].

7.7
Conclusion

The accumulated evidence to date suggests that folate and vitamin B12 play an important role in genomic stability and prevention of DNA hypomethylation in human cells in vitro and in vivo. Above-RDA intakes of these vitamins may be required for a large proportion of humans because of the increasing evidence for common single-nucleotide polymorphisms that significantly alter the activity of proteins required for the absorption, transport, and metabolism of these vitamins to their active forms.

Current evidence from prospective studies suggests a reduced risk of cancer in those with lower chromosomal damage rates, regardless of their exposure to man-made carcinogens. Prevention of DNA hypomethylation is also associated with reduced risk of cancer. We therefore anticipate that increased genomic stability resulting from adequate intake of folate and vitamin B12 may result in reduced risk for cancer. Some epidemiological evidence for this view is already emerging [84, 85, 105, 106].

Other diseases caused by elevated DNA damage, such as infertility, developmental defects, and accelerated aging, may also be prevented by ensuring an adequate intake of folate and vitamin B12.

The importance of choline and methionine for genomic methylation of human cells remains to be defined, but is expected to be an active area of research in the coming decade.

References

1 WAGNER, C. (**1995**) Biochemical role of folate in cellular metabolism. In *Folate in Health and Disease*. Bailey L. B. ed., Marcel Dekker Inc., New York, 23–42.

2 ETO, I.; KRUMDIECK, C. L. (**1986**) Role of vitamin B-12 and folate deficiencies in carcinogenesis. In *Essential nutrients in carcinogenesis*; Poirier, L. A.; Newberne, P. M.; Pariza, M. W.; eds., New York Plenum Press, 313–331.

3 BLOUNT, B. C.; AMES, B. N. (**1995**) DNA damage in folate deficiency, *Bailleres Clin. Haematol. 8*, 461–478.

4 BLOUNT, B. C.; MACK, M. M.; WEHR, C. M.; MACGREGOR, J. T.; HIATT, R. A.; WANG, G.; WICKRAMASINGHE, S. N.; EVERSON, R. B.; AMES, B. N. (**1997**) Folate deficiency causes uracil misincorporation into human DNA and chromosome breakage: implications for cancer and neuronal damage, *Proc. Natl. Acad. Sci. USA 94*, 3290–3295.

5 LINDAHL, T.; WOOD, R.D. (**1999**) Quality control by DNA repair, *Science 286*, 1897–1905

6 ZINGG, J. M.; JONES, P. A. (**1997**) Genetic and epigenetic aspects of DNA methylation on genome expression, evolution, mutation and carcinogenesis, *Carcinogenesis 18*, 869–882.

7 PANCHARUNITI, N.; LEWIS, C. A.; SAUBERLICH, H. E.; PERKINS, L. L.; GO, R.; ALVAREZ, J. O.; MACALUSO, M.; ACTON, R. T.; COPELAND, R. B.; COUSINS, A. L.; GORE, T. B.; CORNWELL, P. E.; ROSEMAN, J. M. (**1994**) Plasma homocysteine, folate and vitamin B-12 concentrations and risk for early-onset coronary artery disease, *Amer. J. Clin. Nutr. 59*, 940–948.

8 FENECH, M. (**2001**) Recommended dietary allowances for genomic stability, *Mutation Res. 480–481*, 51–54.

9 SIBANI, S.; MELNYK, S.; POGRIBNY, I. P.; WANG, W.; HIOU-TIM, F.; DENG, L.; TRASLER, J.; JAMES, S. J.; ROZEN, R. (**2002**). Studies of methionine cycle intermediates (SAM, SAH) DNA methylation and the impact of folate deficiency on tumour numbers in Min mice, *Carcinogenesis 23*, 61–65.

10 GOODMAN, J. E.; LAVIGNE, J. A.; WU, K.; HELZLSOUER, K. J.; STRICKLAND, P. T.; SELHUB, J.; YAGER, J. D, (**2001**) COMT genotype, micronutrients in the folate metabolic pathway and breast cancer risk, *Carcinogenesis 22*, 1661–1665.

11 MATSUO, K.; SUZUKI, R.; HAMAJIMA, N.; OGURA, M.; KAGAMI, Y.; TAJI, H.; KONDOH, E.; MAEDA, S.; ASAKURA, S.; KABA, S.; NAKAMURA, S.; SETO, M.; MORISHIMA, Y.; TAJIMA, K. (**2001**) Association between polymorphisms of folate- and methionine-metabolising enzymes and susceptibility to malignant lymphoma, *Blood 97*, 3205–3209.

12 EPNER, D. E. (**2001**) Can dietary methionine restricition increase the effectiveness of chemotherapy in treatment of advancer cancer? *J. Amer. Coll. Nutr. 20*(5 Suppl), 443S–449 S.

13 LU, S.; EPNER, D. E. (**2000**) Molecular mechanisms of cell cycle block by methionine restriction in human prostate cancer cells, *Nutr. Cancer 38*, 123–130.

14 ZEISEL, S. H. (**2000**) Choline: needed for normal development of memory, *J. Amer. Coll. Nutr. 19*, 528S–531 S.

15 JACOB, R. A.; JENDEN, D. J.; ALLMAN-FARINELLI, M.;A.; SWENDSEID, M. E. (**1999**). Folate nutriture alters choline status of women and men fed low choline diets, *J. Nutr. 129*, 712–717.

16 ZEISEL, S. H.; BLUSZTAJN, J. K. (**1994**) Choline and human nutrition, *Annu. Rev. Nutr. 14*, 269–296.

17 LOCKER, J.; REDDY, T. V.; LOMBARDI, B. (**1986**) DNA methylation and hepatocarcinogenesis in rats fed a choline-devoid diet, *Carcinogenesis 7*, 1309–1312.

18 LOCKER, J.; HUTT, S.; LOMBARDI, B. (**1987**) Alpha-fetoprotein gene methylation and hepatocarcinogenesis in rats fed a choline-devoid diet, *Carcinogenesis 8*, 241–246.

19 ALONSO-APERTE, E.; VARELA-MOREIRAS, G. (**1996**) Brain folates and DNA methylation in rats fed a choline deficient diet with low doses of methotrexate, *Int. J. Vitam. Nutr. Res. 66*, 232–236.

20 TSUJIUCHI, T.; TSUTSUMI, M.; SASAKI, Y.; TAKAHAMA, M.; KONISHI, Y. **(1999)** Hypomethylation of CpG sites and c-myc gene overexpression in hepatocellular carcinomas, but not hyperplastic nodules induced by a choline-deficient L-amino acid-defined diet in rats, *Jpn. J. Cancer Res. 90*, 909–913.

21 BAIK HW and RUSSEL RM (1999) Vitamin B12 deficiency in the elderly. Annu Rev Nutr 19, 357–377.

22 MARSHALL TA, STUMBO PJ, WARREN JJ and XIE XJ (2001) Inadequate nutrient intakes are common and are associated with low diet variety in rural, community-dwelling elderly. J Nutr 131(8), 2192–2196.

23 DWYER JT, MAGAREY AM and DANIELS LA (2001) Evaluation of micronutrient intakes of older Australians: The National Nutrition Survey – 1995. J. Nutr Health Aging 5(4), 243–247.

24 HO C, KAUWELL GP and BAILEY LB (1999) Practitioners' guide to meeting the vitamin B12 recommended dietary allowance for people aged 51 years and older. J Am Diet Assoc 99(6), 725–727.

25 KONINGS, E. J. M.; ROOMANS, H. H. S.; DORANT, E.; GOLDBOHM, R. A.; SARIS, W. H. M.; VAN DEN BRANDT, P. A. **(2001)** Folate intake of the Dutch population according to newly established liquid chromatography data for foods, *Amer. J. Clin. Nutr. 73*, 765–776.

26 FENECH, M.; NOAKES, M.; CLIFTON, P.; TOPPING, D. **(1999)** Aleuorne flour is a rich source of bioavailable folate in humans, *J. Nutr. 129*, 1114–1119.

27 PAUL, A. A.; SOUTHGATE, D. A. T.; RUSSELL, J. **(1987)** McCance and Widdowson's "The Composition of Foods". Her Majesty's Stationery Office, London.

28 SUTHERLAND, G. R. **(1979)** Heritable fragile sites in human chromosomes I. Factors affecting expression in lymphocyte culture, *Amer. J. Hum. Genet. 31*, 125–135.

29 JACKY, P. B.; BEEK, B.; SUTHERLAND, G. R. **(1983)** Fragile sites in chromosomes: possible model for the study of spontaneous chromosome breakage, *Science 220*, 69–70.

30 REIDY, J. A.; ZHOU, X.; CHEN, A. T. L. **(1983)** Folic acid and chromosome breakage I. Implications for genotoxicity studies. *Mutation Res. 122*, 217–221.

31 REIDY, J. A. **(1987)** Folate- and deoxyuridine-sensitive chromatid breakage may result from DNA repair during G2, *Mutation Res. 192*, 217–219.

32 REIDY, J. A. **(1988)** Role of deoxyuridine incorporation and DNA repair in the expression of human chromosomal fragile sites, *Mutation Res. 200(2)*, 215–220.

33 GOULIAN, M.; BLEILE, B.; TSENG, B. Y. **(1980)** Methotrexate-induced misincorporation of uracil into DNA, *Proc. Natl. Acad. Sci. USA 77*, 1956–1960.

34 DUTHIE, S. J.; MCMILLAN, P. **(1997)** Uracil misincorporation in human DNA detected using single cell gel electrophoresis, *Carcinogenesis 18*, 1709–1714.

35 DUTHIE, S. J.; HAWDON, A. **(1998)** DNA instability (strand breakage, uracil misincorporation, and defective repair) is increased by folic acid depletion in human lymphocytes in vitro, *FASEB J. 12*, 1491–1497.

36 CROTT, J. W.; MASHIYAMA, S. T.; AMES, B. N.; FENECH, M. **(2001)** Methylenetetrahydrofolate reductase C677T polymorphism does not alter folic acid deficiency-induced uracil incorporation into primary human lymphocyte DNA in vitro, *Carcinogenesis 22*, 1019–1025.

37 CROTT, J. W.; MASHIYAMA, S. T.; AMES, B. N.; FENECH, M. **(2001)** The effect of folic acid deficiency and MTHFR C677T polymorphism on chromosome damage in human lymphocytes in vitro, *Cancer Epidemiol. Biom. Prev. 10*, 1089–1096.

38 DISCOMBE, G. **(1948)** L'origine des corps de Howell-Jolly et des anneaux de cabot, *Sangre 29*, 262–270.

39 KOYAMA, S. **(1960)** Studies on Howell-Jolly body, *Acta Haemat. Japan 23*, 20–25.

40 LESSIN, L. S.; BESSIS, M. **(1972)** Morphology of the erythron. In *Hematology*, Williams, W. J.; Beutler, E.; Erslev, A. J.; Rundles, R. W. eds., McGraw Hill, New York, 62–93.

41 EVERSON, R. B.; WEHR, C. M.; EREXSON, G. L.; MACGREGOR, J. T. **(1988)** Association of marginal folate depletion with increased human chromosomal damage in vivo: demonstration by analysis of micronucleated erythrocytes, *J. Natl, Cancer Inst, 80*, 525–529.

42 SMITH, D. F.; MacGREGOR, J. T.; HIATT, R. A.; HOOPER, N. K.; WEHR, C. M.; PETERS, B.; GOLDMAN, L. R.; YUAN, L. A.; SMITH, P. A.; BECKER, C. E. (**1990**) Micronucleated erythrocytes as an index of cytogenetic damage in humans: demogrphic and dietary factors associated with micronucleated erythrocytes in splenectomised subjects, *Cancer Res. 50*, 5049–5054.

43 CHEN, A. T. L.; REIDY, J. A.; ANNEST, J. L.; WELTY, T. K.; ZHOU, H. (**1989**) Increased chromosome fragility as a consequence of blood folate levels, smoking status and coffee consumption, *Env. Ml. Mutagenesis 13*, 319–324.

44 MacGREGOR, J. T.; WEHR, C. M.; HIATT, R. A.; PETERES, B.; TUCKER, J. D.; LANGLOIS, R. G.; JACOB, R. A.; JENSEN, R. H.; YAGER. J. W.; SHIGENAGA, M. K.; FREI, B.; EYNON, B. P.; AMES, B.N. (**1997**) "Spontaneous" genetic damage in man: evaluation of interindividual variability, relationship among markers of damage, and influence of nutritional status, *Mutation Res 377*, 125–135

45 TITENKO-HOLLAND, N.; JACOB, R. A.; SHANG, N.; BALARAMAN, A.; SMITH, M. T. (**1998**) Micronuclei in lymphocytes and exfoliated buccal cells of postmenopausal women with dietary changes in folate, *Mutation Res. 417*, 101–114.

46 JACOB, R. A.; GRETZ, D. M.; TAYLOR, P. C.; JAMES, S. J.; POGRIBNY, I. P.; MILLER, B. J.; HENNING, S. M.; SWENDSEID, M. E. (**1998**) Moderate folate depletion increases plasma homocysteine and decreases lymphocyte DNA methylation in postmenopausal women, *J. Nutr. 128*, 1204–1212.

47 PIYATHILKE, C. J.; MACALUSO, M.; HINE, R. J.; VINTER, D. W.; RICHARDS, E. W.; KRUMDIECK, C. L. (**1995**) Cigarette smoking, intracellular vitamin deficiency and occurrence of micronuclei in epithelial cells of the buccal mucosa, *Cancer Epi. Biom. Prev. 4*, 751–758.

48 KRUMDIECK, C. L. (**1991**) Localised folate deficiency and cancer. In *Vitamins and Cancer Prevention*, LAIDLAW, S. A.; SWENDSEID, M. E., eds., Wiley Liss Inc., New York, 39–49

49 KROGH-JENSON, M.; FRIIS-MOLLER, A. (**1967**) Chromosomal studies in pernicious anaemia, *Acta Med. Scand. 181*, 571–576.

50 HEATH, C. W. (**1966**) Cytogenetic observation in vitamin B12 and folate deficiency, *Blood 27*, 800–804.

51 JENSEN, M. K. (**1977**) Cytogenetic findings in pernicious anaemia. Comparison between results obtained with chromosome studies amd the micronucleus test, *Mutation Res. 45*, 249–252.

52 RANA, S. R.; COLMAN, N.; GOH, K.; HERBERT, V.; KLEMPERER, M. R. (**1983**) Transcobalamin II deficiency associated with unusual bone marrow findings and chromosomal abnormalities, *Amer. J. Hematol. 14*, 89–96.

53 WOLLMAN, M. R.; PENCHANSKY, L.; SHEKHTER-LEVIN, S. (**1996**) Transient 7q- in association with megaloblastic anaemia due to dietary folate and vitamin B12 deficiency, *J. Ped. Hem. Oncol. 18*, 162–165.

54 DAS, K. C.; HERBERT, V. (**1978**) The lymphocyte as a marker of past nutritional status: persistence of abnormal lymphocyte deoxyuridine suppression test and chromosomes in patients with past deficiency of folate and vitamin B12, *Br. J. Haematol. 38*, 219–233.

55 GUTTENBACH, M.; SCHMID, M. (**1994**) Exclusion of specific human chromosomes into micronuclei by 5-azacytidine treatment of lymphocyte ceultures, *Exp. Cell Res. 211*, 127–132.

56 XU, G. L.; BESTOR, T. H.; BOURC'HIS, D.; HSIEH, C. L.; TOMMERUP, N.; BUGGE, M.; HULTEN, M.; QU, X.; RUSSO, J. J.; VEIGAS-PEGUINOT, E. (**1999**) Chromosome instability and immunodeficiency syndrome caused by mutations in a DNA methyltransferase gene, *Nature 402*, 187–191.

57 ODAGIRI, Y.; UCHIDA, H. (**1998**) Influence of serum micronutrients on the incidence of kinetochore-positive or -negative micronuclei in human peripheral blood lymphocytes, *Mutation Res. 415*, 35–45.

58 ABRAMSSON-ZETTERBERG, L.; ZETTER-BERG, G.; BERGQVIST, M.; GRAWE, J. (**2000**) Human cytogenetic biomonitoring using flow-cytometric analysis of micrnuclei in transferrin-positive immature peripheral blood reticulocytes. *Env. Mol. Mutagenesis. 36*, 22–31.

59 FENECH. M. (**1993**) The cytokinesis-block micronucleus technique and its application to genotoxicity studies in human populations, *Env. Health Persp. 101*, 101–107.

60 MULLER, W. U.; STREFFER, C. (**1994**) Micronucleus assays. In *Advances in Mutagenesis Research*, Obe, G. ed., Springer-Verlag, London, 1–134.

61 FENECH, M.; RINALDI, J. (**1994**) The relationship between micronuclei in human lymphocytes and plasma levels of vitamin-C, vitamin-E, vitamin B-12 and folic acid, *Carcinogenesis 15*, 1405–1411.

62 FENECH, M.; DREOSTI, I. E.; RINALDI, J. R. (**1997**) Folate, vitamin B12, homocysteine status and chromosome damage rate in lymphocytes of older men, *Carcinogenesis 18*, 1329–1336.

63 FENECH, M.; AITKEN, C.; RINALDI, J. (**1998**) Folate, vitamin B12, homocysteine status and DNA damage in young Australian adults, *Carcinogenesis 19*, 1163–1171.

64 FOWLER, B. M.; GIULIANO, A. R.; PIYATHILAKE, C.; NOUR, M.; HATCH, K. (**1998**) Hypomethylation in cervical tissue – is there correlation with folate status? *Cancer Epi. Biomed. Prev. 7*, 901–906

65 CRAVO, M.; FIDALGO, P.; PEREIRA, A. D.; GOUVEIA-OLIVIERA, A.; CHAVES, P.; SELHUB, J.; MASON, J. B.; MIRA, F. C.; LEITAO, C. N. (**1994**) DNA methylation as an intermediate biomarker in colorectal cancer: modulation by folic acid supplementation, *Eur J. Cancer Prev. 3*, 473–479.

66 BRANDA, R. F.; O'NEILL, J. P.; JACOBSON-KRAM, D.; ALBERTININ, R. J. (**1992**) Factors influencing mutation at the HPRT locus in T-lymphocytes: studies in normal women and women with benign and malignant breast masses, *Env. Mol. Mutagenesis 19*, 274–281.

67 BRANDA, R. F.; O'NEILL, J. P.; SULLIVAN, L. M.; ALBERTININ, R. J. (**1991**) Factors influencing mutation at the HPRT locus in T-lymphocytes:women treated for breast cancer, *Cancer Res. 51*, 6603–6607.

68 CHEN, R. Z.; PETTERSSON, U.; BEARD, C.; JACKSON-GRUSBY, L.; JAENISCH, R. (**1998**) DNA hypomethylation leads to elevated mutation rates, *Nature 395*, 89–93.

69 JONES, P. A. (**1996**) DNA methylation errors and cancer, *Cancer Res. 56*, 2463–2467.

70 POPESCU, N. C.; DIPAOLO, J. A.; AMSBAUGH, S. C. (**1987**) Integration sites of human papillomavirus 18 DNA sequences on HeLa cell chromosomes, *Cytogenet Cell Genet. 44*, 58–62

71 YODER, J. A.; WALSH, C.P.; BESTOR, T. H. (**1997**) Cytosine methylation and the ecology of intragenomic parasites, *Trends Genetics 13*, 335–340.

72 WALSH, C.P.; BESTOR, T. H. (**1999**) Cytosine methylation and mammalian development, *Genes and Development 13*(1), 26–34.

73 WEINBERG, J. B.; SHUGARS, D. C.; SHERMAN, P. A.; SAULS, D. L.; FYTE, A. (**1998**) Cobalamin inhibitors of HIV integrase and integration of HIV1-DNA into cellular DNA, *Biochem. Biophys. Res. Commun. 246*, 393–397.

74 MATHE, G. (**1999**) Why have ten or so nontoxic, retrovirus integrase inhibitors not been made available for AIDS treatment? A ten-year experiment must liberate them. *Biomed. Pharmacotherm 53*, 484–486.

75 CRAVO, M.; GLORIA, L.; CAMILO, M. E.; RESENDE, M.; CARDOSO, J. N.; LEITAO, C. N.; MIRA, F. C. (**1997**) DNA methylation and subclinincal vitamin deficiency of folate, pyridoxal-phosphate and vitamin B12 in chronic alcoholics, *Clin. Nutr. 16*, 29–35.

76 HOMANN, N.; TILLONEN, J.; SALASPURO, M. (**2000**) Microbially produced acetaldehyde from ethanol may increase the risk of colon cancer via folate deficiency, *Int. J. Cancer 86*, 169–173.

77 BAIK, H. W.; RUSSELL, R. M. (**1999**) Vitamin B12 deficiency in the elderly, *Annu. Rev. Nutr. 19*, 357–377

78 BUTZKUEVEN, H.; KING, J. O. (**2000**) Nitrous oxide myelopathy in an abuser of whipped cream bulbs, *J. Clin. Neurosci. 7*, 73–75

79 CARMEL, R. (2000) Current concepts in cobalamin deficiency, *Ann. Rev. Med. 51*, 357–375

80 CHANG, W. P.; LEE, S. R.; TU, J.; HSEU, S. S. (1996) Increased micronucleus formation in nurses with occupational nitrous oxide exposure in operating theatres, *Env. Mol. Mutagenesis 27*, 93–97.

81 FOWLER, B. (1998) Genetic defects of folate and cobalamin metabolism, *Eur. J. Pediatr. 157*, S60–S66.

82 STABLER, S. P.; LINDENBAUM, J.; ALLEN, R. H. (1997) Vitamin B12 deficiency in the elderly: current dilemmas, *Amer. J. Clin. Nutr. 66*, 741–749.

83 ROSENBERG, I. H.; ROSENBERG, L. E. (1998) The implications of genetic diversity for nutrient requirements: the case of folate, *Nutrition Reviews 56*, S47–S53.

84 CHEN, J.; GIOVANNUCCI, E.; KELSEY, K.; RIMM, E. B.; STAMPFER, M. J.; COLDITZ, G. A.; SOEIGELMAN, D.; WILLETT, W. C.; HUNTER, D. J. (1996) A methylenetetrahydrofolate reductase polymorphism and the risk for colorectal cancer, *Cancer Res. 56*, 4862–4864.

85 MA, J.; STAMPFER, M. J.; GIOVANNUCCI, E.; ARTIGAS, C.; HUNTER, D. J.; FUCHS, C.; WILLETT, W. C.; SELHUB, J.; HENNEKENS, C. H.; ROZEN, R. (1997) Methylenetetrahydrofolate reductase polymorphism, dietary interactions and risk of colorectal cancer, *Cancer Res 57*, 1098–1102.

86 JAMES, J. L.; POGRIBNA, M.; POGRIBNY, I. P.; MELNYK, S.; HINE, R. J.; GIBSON, J. B.; YI, P.; TAFOYA, D. L.; SWENSON, D. H.; WILSON, V. L.; GAYLOR, D. W. (1999) Abnormal folate metabolism and mutation in the methylenetetrahydrofolate reductase gene may be maternal risk factors for Down syndrome, *Amer. J. Clin. Nutr. 70*, 495–501.

87 SKIBOLA, C. F.; SMITH, M. T.; KANE, E.; ROMAN, E.; ROLLINSON, S.; CARTWRIGHT, R. A.; MORGAN, G. (1999) Polymorphisms in the methylenetetrahydrofolate reducatse gene are associated with susceptibility to acute leukaemia in adults, *Proc. Nat. Acad. Sci. USA 96*, 12810–12815.

88 AMES, B. N. (1999) Cancer prevention and diet: help from single nucleotide polymorphisms, *Proc. Nat. Acad. Sci. USA 96*, 12216–12218.

89 VAN DER PUT, N. M.; ESKES, T. K.; BLOM, H. J. (1997) Is the common 677C-T mutation in the methylenetetrahydrofolate reductase gene a risk factor for neural tube defects? A meta-analysis, *Q. J. Med. 90*, 111–115.

90 LI, R.; SONIK, A.; STINDL, R.; RASNICK, D.; DUESBERG, P. (2000) Aneuploidy vs. gene mutation hypothesis of cancer: Recent study claims mutation but is found to support aneuploidy. *Proc. Natl. Acad. Sci. USA 97*, 3236–3241.

91 STERN, L. L.; MASON, J. B.; SELHUB, J.; CHOI, S. W. (2000) Genomic DNA hypomethylation, a characteristic of most cancers, is present in peripheral leukocytes of individuals who are homozygous for the C677T polymorphism in the methylenetetrahydrofolate reductase gene, *Cancer Epidemiology Biomarkers Prevention 9*, 849–853.

92 WILSON, A.; PLARR, R.; WU, Q.; LECLERC, D.; CHRISTENSEN, B.; YANG, H.; GRAVEL, R. A.; ROZEN, R. (1999) A common variant in methionine synthase reductase combined with low cobalamin (vitamin B-12) increases risk for spina bifida, *Mol. Genet. Metabol. 67*, 317–323.

93 LACHANCE, P.; LANGSETH, L. (1994) The RDA concept: time for a change, *Nutrition Reviews 52*, 266–270.

94 FENECH, M. (2000) Recommended dietary allowances (RDAs) for genomic stability. *Mutation Research*, Editorial. http://www.mutationresearch.com/ mutat/show/

95 HAGMAR, L.; A. BROGGER, I. HANSTEEN, S. HEIM, B. HOGSTEDT, L. KNUDSEN, B. LAMBERT, K. LINNAINMAA, F. MITELMAN, I. NORDENSON, C. REUTERWALL, S. SALOMAA, S. SKERFVING, M. SORSA (1994) Cancer risk in humans predicted by increased levels of chromosomal aberrations in lymphocytes: Nordic Study Group on the health risk of chromosome damage, *Cancer Res. 54*, 2919–2922.

96 BONASSI, S.; A. ABBONDANDOLO, L. CAMURRI, L. DAL PRA, M. DE FERRARI, F. DEGRASSI, A. FORNI, L. LAMBERTI,

C. LANDO, P. PADOVANI, I. SBRANA, D. VECCHIO, R. PUNTONI (1995) Are chromosome aberrations in circulating lymphocytes predictive of future cancer onset in humans? *Cancer Genet. Cytogenet. 79*, 133–135.

97 BONASSI, S.; HAGMAR, L.; STROMBERG, U.; MONTAGUD, A. H.; TINNERBERG, H.; FORNI, A.; HEIKKILA, P.; WANDERS, S.; WILHARDT, P.; HANSTEEN, I.; KNUDSEN, L. E.; Norrpa, H. for the European Study Group on Cytogenetic Biomarkers and Health (2000) Chromosomal aberrations in lymphocytes predict human cancer independently of exposure to carcinogens, *Cancer Res. 60*, 1619–1625.

98 MIGLIORE, L.; BOTTO, N.; SCARPATO, R.; PETROZZI, L.; CIPRIANI, G.; BONUCCELLI, U. (1999) Preferential occurrence of chromosome 21 malsegregation in peripheral blood lymphocytes of Alzheimer disease patients, *Cytogen. Cell Genet. 87*, 41–46.

99 GUNTER, E. W.; BOWMAN, B. A.; CAUDILL, S. P.; TWITE, D. B.; ADAMS, M. J.; SAMPSON, E. J. (1996) Results of an international round robin for serum and whole-blood folate, *Clin Chem 42*, 1689–1694.

100 DALY, L. E.; KIRKE, P. N.; MOLLOY, A.; WEIR, D. G.; SCOTT, J. M. (1995) Folate levels and neural tube defects, *J. Amer. Med. Assn. 247*, 1698–1702.

101 CUSKELLY, G. J.; McNULTY, H.; SCOTT, J. M. (1996) Effect of increasing dietary folate on red cell folate: implications for prevention of neural tube defects, *Lancet 347*, 657–659.

102 FENECH, M.; NOAKES, M.; CLIFTON, P.; TOPPING, D. (1999) Aleurone flour is a rich source of bioavailable folate, *J. Nutr. 129*, 1114–1119.

103 RIDDELL, L. J.; CHISHOLM, A.; WILLIAMS, S.; MANN, J. I. (2000) Dietary strategies for lowering homocysteine concentrations, *Amer. J. Clin. Nutr. 71*, 1448–1454.

104 WALD, N. J.; BOWER, C. (1994) Folic acid, pernicious anaemia and prevention of neural tube defects, *Lancet 343*, 307.

105 GIOVANNUCCI, E.; STAMPFER, M. J.; COLDITZ, G. A.; HUNTER, D. J.; FUCHS, C.; ROSNER, B. A.; SPEIZER, F. E.; WILLETT, W. C. (1998) Multivitamin use, folate and colon cancer in women in the Nurses Health Study, *Ann. Intern. Med. 129*. 517–524

106 ZHANG, S.; HUNTER, D. J.; HANKINSON, S. E.; GIOVANNUCCI, E. L.; ROSNER, B. A.; COLDITZ, G. A.; SPEIZER, F. E.; WILLETT, W. C. (1999) A prospective study of folate intake and the risk of breast cancer, *J. Amer. Med. Assn. 281*, 1632–1637.

8
Living Longer: The Aging Epigenome

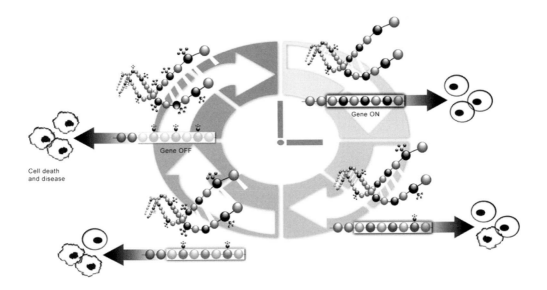

8
Living Longer: The Aging Epigenome

Jean-Pierre Issa*

Summary

The epigenome – a collection of decorations in DNA and associated proteins that dictate permanent states of gene expression – is an increasingly accepted concept in biology with substantial physiological implications. Epigenetic silencing, in particular, is a developmentally critical process that occurs in most multicellular organisms studied to date. A central question regarding the epigenome relates to its stability in adult cells. It is clear that silencing itself is nearly as stable as a genetic change and can generally not be reversed physiologically without passage through the germline. However, it is also apparent that aging cells, particularly epithelial cells, are characterized by an increased frequency of silencing at multiple loci, such that some cells in older individuals have lost the ability to express a certain percentage of previously active genes. The epigenome, then, displays considerable mosaicism in adults, and this mosaicism increases with age and is influenced by lifestyle and environmental exposures. It is proposed that the molecular diversity implied by epigenetic mosaicism contributes to the early stages of acquired, age-related diseases such as neoplasia, cardiovascular disease, and neurodegenerative disease. Measurement of this epigenetic mosaicism may be helpful in determining individual susceptibility to age-related diseases through determination of a "molecular age". Modulation of this process could also lead to advances in the prevention and treatment of age-related diseases, promoting healthier aging and potentially prolonging life.

8.1
Introduction

The genetic revolution of the past decades has largely engrained the concept of immutability of the human genome. This immutability has recently been challenged by recognition of the critical importance of epigenetic modifications in determining

* Acknowledgements: Work in the author's laboratory is supported by grants from the National Institutes of Health, USA, and the George and Barbara Bush Endowment for Innovative Cancer Research.

final gene-expression patterns [1, 2] Epigenetic patterns (the "epigenome") are significantly more fluid than genetic patterns, and their potential modulation by lifestyle and environmental exposures adds new molecular parameters to the old debates on nature vs. nurture [3]. Indeed, the clonal nature of such exposure-related epigenetic changes, and their potential in some organisms to be transmitted to subsequent generations, have even raised discussions of non-Mendelian inheritance, a mostly taboo subject during the unraveling of the genetic code. In this review, epigenetic changes during human aging are described and proposed to lead to substantial epigenetic mosaicism in older individuals. Such epigenetic mosaicism provides fertile ground for the acquisition of additional genetic and epigenetic alterations that contribute to age-related diseases.

8.2
Methylation as a Mark of Epigenetic Silencing

DNA methylation is an enzymatic modification that, in mammals, affects primarily the cytosine base when it is followed by a guanosine (the CpG dinucleotide) [4]. Methylation in CpG-rich areas termed CpG islands is associated with epigenetic silencing of the involved genes [4]. This silencing is stable, mitotically transmitted, and, in most cases, physiologically irreversible, short of passage through the germline. In humans, methylation-associated silencing is best exemplified by X inactivation in women [5], where one X-chromosome is randomly targeted for transcriptional silencing, and by imprinting, where a single allele (or promoter) of a given gene is silenced on the basis of parental origin [6]. Dense promoter methylation often accompanies irreversible loss of expression in these cases.

Methylation has been unequivocally linked to maintenance of the silent state [4]. Disruption of methylation pharmacologically or genetically results in reactivation of gene expression at previously solidly silent loci [7]. Plausible mechanisms for methylation-associated silencing have been advanced, including recruitment of methyl-binding proteins and associated protein complexes that remodel chromatin locally [8]. In addition, methyltransferases themselves can achieve some degree of transcriptional suppression, partly through the recruitment of histone-modifying enzymes [9]. Nevertheless, there remains some doubt as to whether methylation itself initiates the silencing process or whether transcriptional suppression allows methylation to settle in a given promoter and maintain the silent state [10]. Whatever the final resolution of this question reveals, it is clear that promoter methylation provides at least a marker of transcriptional silencing that can be relatively easily detected and quantified.

8.3
The Dogma: Formation of Methylation Patterns

Adult patterns of methylation are efficiently erased early in embryogenesis via both active and passive mechanisms [11]. From this clean slate, methylation patterns are reestablished soon after implantation, following complex and poorly understood mechanisms. The pioneering work of Turker et al. [12] suggested that methylation is targeted to specific areas of the genome via "methylation centers" which, through unknown molecular mechanisms, attract DNA methyltransferases efficiently. This nidus of methylation then spreads in the 5′ and 3′ directions relatively efficiently during embryogenesis, to distances of several hundred base pairs. Methylation spreading, particularly to promoter regions, is opposed by trans-activating factors that provide protection against methylation-associated silencing [13, 14]. Methylation spreading appears to be one of the central properties of interactions between DNA methyltransferase and DNA. Besides occurring in native DNA, spreading also occurs in integrated transgenes and viral sequences [15]. As discussed later, spreading can also be observed in neoplastic and aging methylation. The molecular nature of these processes of spreading and protection is as poorly understood as the nature of methylation centers. In addition, the contribution of demethylases [16] to final methylation patterns remains to be clarified.

Once formed post-implantation, methylation patterns have been assumed to remain immutable. Thus, a complex picture emerged whereby promoter-associated CpG islands are unmethylated regardless of gene expression, and CpG-poor promoters are generally unmethylated in expressing tissues and methylated in non-expressing tissues [17]. Exceptions to these rules included primarily the genes epigenetically silenced by imprinting or X-chromosome inactivation. Recently, a class of genes with promoters of intermediate CpG content has been described, with evidence of tissue-specific methylation patterns [18]. In all these cases, however, classical models of methylation patterning imparted a certain finality to the distribution of 5-methylcytosine in adult cells.

8.4
The Reality: Age-related Methylation and Epigenetic Mosaicism in Normal Epithelium

De-novo methylation of CpG island-associated promoters is a common event in human neoplasia [19], estimated to affect several hundred genes in a given tumor [20]. The origin of this aberrant methylation has been debated for years. It is now clear that, in colon cancer, the vast majority of aberrant methylation in cancerous tissue can also be found to a smaller degree in normal tissues [21], including tissues geographically distant from the neoplasm [22]. Such methylation in normal tissues is not limited to patients with cancer. In fact, it occurs as an age-related trend that is linear across the population but shows substantial variability within a given age group [23]. By extrapolation from the cancer situation, it is likely that hundreds of genes are affected by age-related methylation in a given individual. A partial list of genes showing promoter methylation with age in normal tissues is shown in Table 8.1.

What are the functional consequences of age-related methylation? The promoter regions studied correspond to areas where methylation would be expected to silence the genes. Indeed, in cancers, where methylation is dense, silencing of these regions can be easily demonstrated, along with reactivation by methylation inhibition [24].

Tab. 8.1 Some genes displaying promoter and CpG-island methylation in apparently normal aging tissues; Xs refers to chromosomal location

Gene	Xs	Tissue	References
ERα	6q25.1	Colon, liver, heart, fibroblasts	[19, 22, 40, 50]
CSPG2	5q12-14	Colon	[24]
MLH1	2p22	Colon	[27]
E-cadherin	16q22.1	Bladder	[51]
MYOD	11p15.4	Colon	[28]
APC	5q21	Stomach	[30]
N33	8p22	Colon	[28]
HIC1	17p13.3	Prostate	Issa, JP, unpublished
IGF2	11p15.5	Colon, fibroblasts	[31]
PAX6	11p13	Colon	[32]
RARβ2	3p24	Colon	Issa, JP, unpublished
DBCCR1	9q32-33	Bladder	[33]
COL1A1	17q21-22	Periodontal ligament	[34]

However, age-related methylation is typically partial, with quantitative assays revealing ranges of 5%–50%, although most values are in the lower end of this range [21]. Such low levels of methylation can be interpreted in two ways: methylation could be sparsely affecting all cells under study, which may have little effect on gene expression, or methylation could be extensive, but limited to a few cells in a given population. In the colon, there is evidence for both situations, depending on the gene. A study of a transgene specific to colonic epithelial cells showed that age-related silencing was uniform across individual crypts but mosaic across the colon [25]. Thus, individual crypts, derived from individual stem cells, were autonomously affected by silencing, and the total number of silenced crypts increased substantially with age. Given that transgene silencing often mimics methylation-associated silencing, one would hypothesize that age-related methylation could also be crypt (and stem cell) autonomous, with considerable crypt-to-crypt variability. Indeed, a recent study using microdissected crypts and the bisulfite genomic sequencing method revealed such crypt-to-crypt heterogeneity [26], with some densely methylated crypts where silencing would be expected, and completely unmethylated crypts that, presumably, retained the capacity to express the target genes. Bisulfite-sequencing of non-microdissected colonic tissues also reveals the presence of molecules with complete methylation, presumably reflecting crypts where the gene has been completely silenced [27].

What is the cause of age-related methylation? The general patterns one observes are most consistent with an extension of the development-related establishment of methylation patterns [12]. Thus, methylation centers appear to exist that initially attract age-related methylation, with progressive spreading towards the promoter area. For example, MYOD methylation has been reported to spread from exonic regions to the promoter region with age [28]. A study of intermediately methylated MLH1 alleles suggests that methylation spreads from a nidus about 1 kb upstream of the transcription start [27]. It was proposed [29] that age-related methylation reflects continuing low-efficiency spreading of methylation after the initial wave of de-novo methylation observed in embryogenesis. The rate of methylation, however, is likely to be substantially variable among different tissues and different individuals [28], and the sources of this variability are of considerable biological and clinical interest.

8.5
The Consequences: Methylation and Age-related Diseases

Does age-related epigenetic mosaicism have any physiological consequence? At the tissue level, it is clear that gene expression changes that can be accounted for by this process are modest, given the relatively minor degree of age-related methylation observed [22]. Indeed, for some genes, this phenomenon can be observed only with very sensitive methods [27]. Arguably, then, methylation cannot be a major part of age-related gene expression changes observed using microarray technology [30] or of substantial declines in normal tissue function such as loss of elasticity [31]. However, age-related diseases are often related to focal changes that conceivably begin in a single cell, as in neoplastic proliferation or atherosclerotic plaque formation. These focal changes may well be related to epigenetic mosaicism. Indeed, the diversity in individual cell phenotypes that can be achieved by epigenetic mechanisms likely promotes the development of processes that involve clonal selection, such as neoplasia.

A role for epigenetic mechanisms in cancer development is now recognized. Global chromatin and transcriptional regulators are genetically altered in a variety of human neoplasms [32–34] and downstream epigenetic alterations are likely key to the physiologic effects of such genetic changes. In addition, a more direct role for epigenetic alterations in cancer is suggested by the frequent occurrence of transcriptional silencing in association with promoter methylation in many human tumors [19, 35]. This silencing is clearly an alternative mechanism for pathway inactivation in cancer, and affected genes (including tumor-suppressor genes and DNA-repair genes) are critical mediators of the neoplastic phenotype. Remarkably, most careful studies have demonstrated that methylation-type epigenetic defects are often found in preneoplastic lesions [21, 36, 37] and in normal-appearing tissues of patients at risk for neoplasia [29]. It is clear that age-related epigenetic mosaicism could play a major role in the development of human neoplasms by generating the underlying molecular diversity required for the early steps of malignant transformation. In other words, age-related changes could predispose a small fraction of normal epithelium to neoplastic transformation by subtly altering the expression of key genes, such as *MLH1* in the colon. Although these changes involve a minor fraction of normal tissues, the positive selection afforded by increased proliferation or decreased apoptosis explains their prominence in neoplastic lesions. The interactions between epigenetic mosaicism recapitulate, to a certain extent, the processes at work in Darwinian evolution: Molecular diversity at the cellular level provides the selective engine for survival of the fittest – in this case, unfortunately, these are neoplastic cells.

Another disease characterized by focal proliferative lesions is atherosclerotic vascular disease [38]. Uncontrolled smooth-muscle cell proliferation contributes to narrowing of the vascular lumen and, eventually, to cardiac or neurologic ischemia.

This proliferation is clearly multifactorial, but resembles some aspects of the uncontrolled growth of neoplastic cells [38]. A role for methylation-mediated epigenetic defects in the process was suggested by studies of the estrogen receptor alpha gene (*ERα*). *ERα* expression is often lost in aging and diseased blood vessels [39], and age-related methylation of its promoter has been demonstrated in cardiac tissues [40], as well as in atherosclerotic lesions and cultured vascular smooth-muscle cells [40, 41]. Ongoing studies should clarify whether this process affects other genes in vascular tissues. In principle, age-related epigenetic mosaicism in vascular endothelial and smooth-muscle cells could help promote atherosclerosis in a manner analogous to its promotion of neoplastic transformation.

Other acquired and age-related diseases may also be affected by promoter methylation in normal tissues, with its associated epigenetic alterations. These include common disorders such as senile dementia, where focal cell dysfunction contributes to the pathophysiology, and insulin resistance and associated diabetes, where signal transduction through the insulin receptor appears to be progressively impaired with age. In addition, hypermethylation has now been demonstrated in multiple inflammatory states such as inflammatory bowel disease [42], Barrett's esophagus [43], and chronic hepatitis [44], and it is likely that "accelerated" silencing plays a role in the tissue dysfunction observed in some such states. The contribution of epigenetic defects to non-neoplastic diseases is only beginning to be studied and may well reveal surprises in the coming few years.

8.6
The Bottom Line: Clinical Implications

Why does all this matter? Beyond revisiting some aspects of the pathogenesis of age-related diseases, this dynamic view of the epigenome has substantial implications for disease risk assessment, prevention, and treatment [7]. We are just beginning to explore in earnest the clinical potential of the aging epigenome concept.

As mentioned above, it is clear that the rate of age-related methylation substantially varies among healthy individuals and in patients with certain diseases such as inflammatory bowel disease. Measurement of methylation variation may thus more precisely estimate risk of developing various diseases than population estimates, by providing an approximation of physiologic age and the exposures that modify this process. For example, measurement of methylation could provide an estimate of disease activity and risk of cancer development in inflammatory bowel disease, chronic hepatitis, or Barrett's esophagus. Substantial hurdles need to be overcome before this concept becomes reality, including identification of physiologically relevant methylation markers, validation of their potential as discriminatory markers, and relatively large epidemiological studies to confirm the findings. Nevertheless, this approach could have important practical implications by reducing the guesswork involved in estimating disease risk. Eventually, large-scale methylation profiling of normal tissues could theoretically have profound public-health implications by creating a risk stratification for common diseases (cancer, coronary artery disease, neurodegenerative diseases, etc.), as well as by identifying lifestyle and exposure factors that affect disease risk via epigenetic modifications.

Beyond disease-risk assessment, the real public-health potential of understanding age-related epigenetic mosaicism is the possibility of intervening in the process for preventive purposes. In animal models, it is clear that reducing DNA methyltransferase levels can affect the risk of tumor development [45]. In humans, this field is still in its infancy. Emerging data point to specific lifestyle and environmental exposures that accelerate the process (Issa et. al., unpublished observations), and modification of these risk factors will likely affect the degree of age-related methylation and its associated pathology. Epidemiological studies should lead to subtler and more common factors that may, in fact, slow the process and affect disease in that way. For example, there is considerable current interest in the possibility of interactions between diet and DNA methylation [53], and ongoing studies should help clarify the potential of diet in reducing age-related methylation.

Finally, although disease prevention is a worthwhile goal, it is the prospect of epigenetic intervention that fuels much of the current excitement in the field [7]. Unlike genetic changes, which are largely irrevocable once established (at least using current technology), epigenetic changes have the potential to be reversible, given that gene function is altered by transcriptional mechanisms rather than via protein structural changes. The ultimate proof of the reversibility of this process lies in the process of epigenetic programming during embryogenesis, whereby epigenetic patterns are nearly completely erased, and the embryo starts with a clean slate [46].

Can epigenetic reprogramming be achieved in adult cells? Clearly, nuclear transfer technology can result in substantial epigenetic reprogramming, albeit at very low efficiency [47]. However, there is evidence that pharmacological manipulation of epigenetic processes does in fact have the potential to also achieve some degree of reprogramming. This was convincingly demonstrated two decades ago when fibroblasts were induced to differentiate into muscle cells by using inhibitors of DNA methyltransferases [48]. This phenomenon was subsequently shown to be related to the demethylation and activation of a silenced muscle-specific transcription factor, MYOD1 [49]. DNA-methyltransferase inhibitors can reverse age- and cancer-related silencing in vitro [24] and, in preliminary studies, in vivo as well [58]. It is hoped that the clinical development of epigenetic therapy will eventually affect age-related diseases and afford some degree of life prolongation, particularly in individuals at risk for such diseases.

References

1 A. P. WOLFFE, M. A. MATZKE, *Science* **1999**, *286*, 481–486.

2 T. JENUWEIN, C. D. ALLIS, *Science* **2001**, *293*, 1074–1080.

3 R. L. JIRTLE, M. SANDER, J. C. BARRETT, *Environ. Health Perspect.* **2000**, *108*, 271–278.

4 P. A. JONES, D. TAKAI, *Science* **2001**, *293*, 1068–1070.

5 S. M. GARTLER, M. A. GOLDMAN, *Curr. Opin. Pediatr.* **2001**, *13*, 340–345.

6 D. P. BARLOW, *Science* **1995**, *270*, 1610–1613.

7 V. SANTINI, H. M. KANTARJIAN,

J. P. Issa, *Ann. Intern. Med.* **2001**, *134*, 573–586.

8 A. P. Bird, A. P. Wolffe, *Cell* **1999**, *99*, 451–454.

9 F. Fuks, W. A. Burgers, A. Brehm, L. Hughes-Davies, T. Kouzarides, *Nat. Genet.* **2000**, *24*, 88–91.

10 J. Z. Song, C. Stirzaker, J. Harrison, J. R. Melki, S. J. Clark, *Oncogene* **2002**, *21*, 1048–1061.

11 A. Razin, R. Shemer, *Hum. Mol. Genet.* **1995**, 4 Spec No, 1751–1755.

12 M. S. Turker, *Semin. Cancer Biol.* **1999**, *9*, 329–337.

13 D. Macleod, J. Charlton, J. Mullins, A. P. Bird, *Genes Dev* **1994**, *8*, 2282–2292.

14 M. Brandeis, D. Frank, I. Keshet, Z. Siegfried, M. Mendelsohn, A. Nemes, V. Temper, A. Razin, H. Cedar, *Nature* **1994**, *371*, 435–438.

15 W. Doerfler, U. Hohlweg, K. Muller, R. Remus, H. Heller, J. Hertz, *Ann. N. Y. Acad. Sci.* **2001**, *945*, 276–288.

16 S. K. Bhattacharya, S. Ramchandani, N. Cervoni, M. Szyf, *Nature* **1999**, *397*, 579–583.

17 M. S. Turker, T. H. Bestor, *Mutat. Res.* **1997**, *386*, 119–130.

18 C. De Smet, C. Lurquin, B. Lethe, V. Martelange, T. Boon, *Mol. Cell Biol.* *19*, 7327–7335.

19 S. B. Baylin, J. G. Herman, J. R. Graff, P. M. Vertino, J. P. J. Issa, *Adv. Cancer Res.* **1998**, *72*, 141–196.

20 J. F. Costello, M. C. Fruhwald, D. J. Smiraglia, L. J. Rush, G. P. Robertson, X. Gao, F. A. Wright, J. D. Feramisco, P. Peltomaki, J. C. Lang, D. E. Schuller, L. Yu, C. D. Bloomfield, M. A. Caligiuri, A. Yates, R. Nishikawa, H. H. Su, N. J. Petrelli, X. Zhang, M. S. O'Dorisio, W. A. Held, W. K. Cavenee, C. Plass, *Nat. Genet.* **2000**, *24*, 132–138.

21 M. Toyota, N. Ahuja, M. Ohe-Toyota, J. G. Herman, S. B. Baylin, J. P. J. Issa, *Proc. Natl. Acad. Sci. US A* **1999**, *96*, 8681–8686.

22 J. P. Issa, Y. L. Ottaviano, P. Celano, S. R. Hamilton, N. E. Davidson, S. B. Baylin, *Nat. Genet.* **1994**, *7*, 536–540.

23 J. P. Issa, *Curr. Top. Microbiol. Immunol.* **2000**, *249*, 101–118.

24 W. S. Post, P. J. Goldschmidt-Clermont, C. C. Wilhide, A. W. Heldman, M. S. Sussman, P. Ouyang, E. E. Milliken, J. P. Issa, *Cardiovasc. Res.* **1999**, *43*, 985–991.

25 J. P. Issa, P. M. Vertino, J. Wu, S. Sazawal, P. Celano, B. D. Nelkin, S. R. Hamilton, S. B. Baylin, *J. Natl. Cancer Inst.* **1993**, *85*, 1235–1240.

26 M. Toyota, C. Ho, N. Ahuja, K.-W. Jair, M. Ohe-Toyota, S. B. Baylin, J. P. J. Issa, *Cancer Res.* **1999**, *59*, 2307–2312.

27 H. Nakagawa, G. J. Nuovo, E. E. Zervos, E. W. Martin, Jr., R. Salovaara, L. A. Aaltonen, C. A. de la Chapelle, *Cancer Res.* **2001**, *61*, 6991–6995.

28 D. M. Bornman, S. Mathew, J. Alsruhe, J. G. Herman, E. Gabrielson, *Amer. J. Pathol.* **2001**, *159*, 831–835.

29 N. Ahuja, Q. Li, A. L. Mohan, S. B. Baylin, J. P. Issa, *Cancer Res.* **1998**, *58*, 5489–5494.

30 T. Tsuchiya, G. Tamura, K. Sato, Y. Endoh, K. Sakata, Z. Jin, T. Motoyama, O. Usuba, W. Kimura, S. Nishizuka, K. T. Wilson, S. P. James, J. Yin, A. S. Fleisher, T. Zou, S. G. Silverberg, D. Kong, S. J. Meltzer, *Oncogene* **2000**, *19*, 3642–3646.

31 J. P. J. Issa, P. M. Vertino, C. D. Boehm, I. F. Newsham, S. B. Baylin, *Proc. Natl. Acad. Sci. USA* **1996**, *93*, 11757–11762.

32 M. Toyota, J. P. Issa, *Semin. Cancer Biol.* **1999**, *9*, 349–357.

33 T. Habuchi, T. Takahashi, H. Kakinuma, L. Wang, N. Tsuchiya, S. Satoh, T. Akao, K. Sato, O. Ogawa, M. A. Knowles, T. Kato, *Oncogene* **2001**, *20*, 531–537.

34 M. Takatsu, S. Uyeno, J. Komura, M. Watanabe, T. Ono, *Mech. Ageing Dev.* **1999**, *110*, 37–48.

35 S. M. Cohn, K. A. Roth, E. H. Birkenmeier, J. I. Gordon, *Proc. Natl. Acad. Sci. USA* **1991**, *88*, 1034–1038.

36 Y. Yatabe, S. Tavare, D. Shibata, *Proc. Natl. Acad. Sci. USA* **2001**, *98*, 10839–10844.

37 J. P. Issa, *Ann. N. Y. Acad. Sci.* **2000**, *910*, 140–153.

38 C. K. Lee, R. G. Klopp, R. Weindruch,

T. A. Prolla, *Science* **1999**, *285*, 1390–1393.

39 D. Harman, *Ann. N. Y. Acad. Sci.* **2001**, *928*, 1–21.

40 I. Versteege, N. Sevenet, J. Lange, M. F. Rousseau-Merck, P. Ambros, R. Handgretinger, A. Aurias, O. Delattre, *Nature* **1998**, *394*, 203–206.

41 M. van Lohuizen, *Curr. Opin. Genet. Dev.* **1999**, *9*, 355–361.

42 R. H. Giles, D. J. Peters, M. H. Breuning, *Trends Genet.* **1998**, *14*, 178–183.

43 P. A. Jones, P. W. Laird, *Nat. Genet.* **1999**, *21*, 163–167.

44 M. Toyota, N. Ahuja, H. Suzuki, F. Itoh, M. Ohe-Toyota, K. Imai, S. B. Baylin, J. P. Issa, *Cancer Res.* **1999**, *59*, 5438–5442.

45 S. Zochbauer-Muller, K. M. Fong, A. K. Virmani, J. Geradts, A. F. Gazdar, J. D. Minna, *Cancer Res.* **2001**, *61*, 249–255.

46 J. S. Ross, N. E. Stagliano, M. J. Donovan, R. E. Breitbart, G. S. Ginsburg, *Ann. N. Y. Acad. Sci.* **2001**, *947*, 271–292.

47 D. W. Losordo, M. Kearney, E. A. Kim, J. Jekanowski, J. M. Isner, *Circulation* **1994**, *89*, 1501–1510.

48 A. K. Ying, H. H. Hassanain, C. M. Roos, D. J. Smiraglia, J. J. Issa, R. E. Michler, M. Caligiuri, C. Plass, P. J. Goldschmidt-Clermont, *Cardiovasc. Res.* **2000**, *46*, 172–179.

49 J. P. Issa, N. Ahuja, M. Toyota, M. P. Bronner, T. A. Brentnall, *Cancer Res.* **2001**, *61*, 3573–3577.

50 C. A. Eads, R. V. Lord, S. K. Kurumboor, K. Wickramasinghe, M. L. Skinner, T. I. Long, J. H. Peters, T. R. DeMeester, K. D. Danenberg, P. V. Danenberg, P. W. Laird, K. A. Skinner, *Cancer Res.* **2000**, *60*, 5021–5026.

51 Y. Kondo, Y. Kanai, M. Sakamoto, M. Mizokami, R. Ueda, S. Hirohashi, *Hepatology* **2000**, *32*, 970–979.

52 P. W. Laird, L. Jackson-Grusby, A. Fazeli, S. L. Dickinson, W. E. Jung, E. Li, R. A. Weinberg, R. Jaenisch, *Cell* **1995**, *81*, 197–205.

53 Issa JPJ, J. Nutr., In press, 2002 MAT-Marjorie A. TiefertRef.23: need more complete citation.

54 W. Reik, W. Dean, J. Walter, *Science* **2001**, *293*, 1089–1093.

55 W. M. Rideout III, K. Eggan, R. Jaenisch, *Science* **2001**, *293*, 1093–1098.

56 P. G. Constantinides, P. A. Jones, W. Gevers, *Nature* **1977**, *267*, 364–366.

57 P. A. Jones, M. J. Wolkowicz, W. M. Rideout III, F. A. Gonzales, C. M. Marziasz, G. A. Coetzee, S. J. Tapscott, *Proc. Natl. Acad. Sci. USA* **1990**, *87*, 6117–6121.

58 R. K. Mannari, H. M. Kantarijian, G. Garcia-Manero, J. Cortes, M. Beran, J. P. Issa, *Proc. AACR* **2001**, *42*, 709.

9
Digitizing Molecular Diagnostics: Current and Future Applications of Epigenome Technology

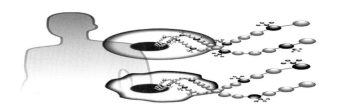

All cells shed DNA into the bloodstream

9

Digitizing Molecular Diagnostics: Current and Future Applications of Epigenome Technology

SVEN OLEK, SABINE MAIER, KLAUS OLEK, and ALEXANDER OLEK

Summary

Methylation has enjoyed growing attention by the scientific community during the past decade. However, it has been regarded as a rather exotic biological phenomenon, and few have actually begun thinking about, let alone investing in, actual applications of this exciting science. In this chapter we argue that methylation has the general characteristics needed for an ideal diagnostic testing technology, applicable to most common (noninfectious) diseases. Specifically, we argue that a methylation-based strategy would be the perfect (and maybe only) tool for creating a comprehensive cancer-management system, integrating detection in asymptomatic people, monitoring, cancer resistance, and pharmacogenetic testing. In addition, and just as an example for a variety of other applications, we discuss the general advantages of methylation as a phenotyping tool for quality control of tissue engineered organ replacements, as well as process control in tissue engineering. Last, and not least, we illustrate applications in animal breeding, as one instance of an opportunity in agriculture.

9.1
Introduction

As we have learned in the previous chapters, methylation plays a key role in several biological processes. It also correlates with – or may even be the causative factor for – several common, serious diseases. Both as a diagnostic parameter, and as a means of furthering our understanding of disease (and thereby contributing to the development of new therapies), methylation has particular advantages over other state-of-the-art technologies. Identifying those areas in which methylation has unique potential and developing products to apply methylation in diagnostics, pharmaceutical research, tissue engineering, and even agricultural research, is part of the mission of the companies that were founded or cofounded by the authors of this chapter.

In this chapter, we discuss some possible "future applications" of methylation, a list that may, however, not be complete. Some of the discussed opportunities are already in advanced product development at our companies Epigenomics, Epiontis ("Epigenetics On Tissues"), Biopsytec, and their partners and should become available for applications within the next few years. Others are truly future prospects, the promises of which are based on circumstantial evidence, fewer and basic experiments only, or data that are not publicly available. Hence, we discuss opportunities in the order in which we expect their application to become available for actual product development, leaving the basic research phase. We intend to present the most transparent picture of our view of future applications, backed by several years of our own commercial research in the area, in addition to published data.

9.2
Molecular Diagnostics

9.2.1
Advantages of Methylation as a Diagnostic Parameter

Diagnostics today are usually based on either of two approaches. Ad-hoc testing of whether a patient has a disease, and if so, what form of it, is the topic of classical diagnostics as used over much of the second half of the last century. The methods used are manifold and include imaging, in-vivo diagnostics, protein-based ELISA assays, cell counts, and many more. Future diagnostics will be different from today's in several aspects:

- Most probably, one of the main characteristics of future diagnostics is that they will be multiparametric, i.e., several molecular markers will be analyzed and taken into account synergistically to answer single, but complicated, diagnostic questions. Results from individual markers will be fed into algorithms that have been "trained" in clinical studies, to yield compound results in which the total knowledge gained is greater than the sum of the parts.
- Diagnostic markers will often be based on nucleic acids, allowing more systematic development of new tests.
- Diagnostics will be used for disease management, successively answering different questions for the same patient

For several reasons, this change from what we call the single-marker paradigm has not yet truly begun. Some reasons relate to the information content of the molecules (i.e., DNA, RNA, or proteins) that are being used for diagnostics; others relate to technical characteristics of these molecules (e.g., the instability of RNA) or to the difficulty regulatory authorities have in approving new tests that are based on a new paradigm rather than on incremental additions to an existing one.

Methylation is the ideal parameter for comprehensive diagnostics in disease management, with the potential to contribute to or even trigger a change in paradigm from single to multiple markers and from individual questions to a disease-management approach.

Two sets of reasons make analysis of methylation patterns the ideal diagnostic tool. One relates to the special information content of methylation patterns, the other to clear-cut technical advantages that will make the design and operation of testing devices and platforms easier and more reliable.

Information content of methylation patterns. DNA methylation has several characteristics that make it a useful source of information for application in diagnostic products:

- Some methylation positions in a gene provide information about its current activation state [1].
- Other positions in the promoter of a gene can provide information as to how easily it can be activated, i.e., whether the gene is "ready to go" – allowing one to predict the reaction of a cell to stimuli such as drugs [2–6].
- The methylation pattern of the sequences to which transcription factors bind allows predictions about the binding of transcription factors or other chromatin components [1].
- Expression of genetic networks is often orchestrated through chromatin remodeling. Since this is reflected or even controlled by methylation, alterations of many genes can be tested from only few genomic sites [1, 6].
- The methylation patterns of cells greatly vary with age of an individual. As these alterations occur in gene-regulatory sequences, age-dependent dysregulation of gene expression through methylation changes will lead to disease [4, 8, 9].
- Nutritional and environmental influences, such as uptake of toxins, have a significant effect on the genomic methylation pattern. Hence, poor nutrition could cause regulatory disturbances in the genome, leading to metabolic disorders or cardiovascular disease [10, 11].
- Since the methylation pattern is heritable through cell divisions, past events, such as environmental influences during development, can be reflected [14].
- Methylation patterns at many positions in the genome differ among the tissues of each individual [12]

Technical advantages of methylation-based platforms. Diagnostic information needs to be obtained both on parameters related to genetic constitution and predispositions of patients, as well as on processes actually going on in affected cells or tissues, i.e., gene expression.

For example, the cytochrome family of drug-metabolizing enzymes plays a ubiquitous role in the response of patients to drugs. The metabolizing profile of cytochromes underlies significant variation caused by the existence of many different alleles in the population. However, more recently cytochromes were found to be also heavily expression regulated, notably in connection with changes in methylation patterns [4]. Up-regulation of enzymes can compensate for the presence of weaker alleles, so that in borderline cases the benefit of knowing the identity (and function) of the alleles in a person would have no value whatsoever. Hence, to reliably obtain a notion of how effectively a patient will metabolize a drug (and in consequence, whether he or she is likely to be affected by adverse side effects), gene-expression and genetic profiles have to be generated.

Cost is a major factor in determining whether a diagnostic product is reimbursed by insurance companies, and therefore whether it is likely to be widely adopted. Methylation analysis can yield a good approximation of gene activity, as well as genetic (single nucleotide polymorphism, SNP) information within one and the same assay, using one and the same template molecule.

Several additional technical advantages make methylation an ideal tool for routine diagnostics assays:

- DNA is a very stable molecule. Unlike mRNA and proteins, DNA samples can be handled without any particular precautions under normal clinical and laboratory conditions.
- Using DNA as the analyte, methylation and SNP signals can be measured in the same systems, providing the opportunity to see dynamic gene regulation and predisposition information on the same technological platform with no additional effort. Assaying methylation therefore is the only way to create a true one-stop shop for both fundamental types of biological information, the program (DNA) and its execution (mRNA and proteins).
- Methylation information relevant to tumors can be acquired from either circulating tumor cells or free DNA in the blood of cancer patients.
- The sensitivity of measuring methylation is very high.
- Methylation signals are digital: each cytosine on a chromosome can only be in "on" (methylated) or "off" (not methylated).
- Repeated PCR-amplification cycles do not disturb the sharpness of the methylation signal, so it can be performed for large numbers of low- and high-copy genes at the same time with minute sample sizes.
- These features of DNA methylation allow indexing of signals and results through reliable calibration. The lack of this possibility in the competing expression-profiling strategy is so obvious, that even an international effort to create an industry standard for mRNA profiling has not yet been successful.
- The ability to analyze methylation signals from paraffin-, formalin-, or alcohol-preserved samples allows expression analysis of archived samples, making possible the exploitation of the vast treasure of medical data stored by pathologists.
- The analysis of methylation can be performed equally well by DNA microarrays, real-time PCR, MALDI mass spectrometry, or any other DNA analysis technology, allowing use of the most efficient methods for various possible applications. In particular, different technologies can be employed to solve problems in diverse research settings.

9.2.2
Cancer Management

Early detection. In the future, cancer will be systematically diagnosed much earlier than today. This fact alone has the potential for prolonging the lives of millions. The search for informative protein markers for early diagnosis of cancer has been going on for decades, yielding only a few major breakthroughs. And, although proteomics promises to accelerate the search for protein markers, some fundamental disadvantages compared to nucleic acid-based techniques will remain. For example, directed and efficient whole-genome subtractive techniques are not an option in proteomics. Technical shortcomings of the actual test platforms (such as insufficient sensitivity due to our inability to amplify proteins) remain a significant hurdle. Therefore, the alternative concept of serum- or plasma-based nucleic-acid screening is appealing.

Due to the stability of DNA, as well as the potentially "binary" nature of methylation signals (an "on" signal indicative of a tumor is amplified against a background of "off" signals originating from healthy cells), methylation is optimally suited for early detection of cancer in blood serum or plasma. It is thought that those nucleic acid fragments that are wrapped around nucleosomes remain stable in serum, giving DNA yet another advantage over RNA as the diagnostic analyte.

Ample evidence exists for elevated DNA (and, incidentally, elevated nucleosome) concentrations in the blood of cancer patients (Epigenomics, unpublished data, [(40, 41]). Several markers have been found that are methylated only in DNA that is shed by cancerous cells. These fundamental pre-requisites, shown by many academic groups, demonstrated the proof-concept for this approach to cancer screening. More recently, actual product development has begun at Epigenomics, and the first methylation-based early-detection methods will soon enter clinical trials. This application of methylation ought to lead to a major improvement in survival and much less suffering (e. g., through less damaging surgeries) for millions of cancer patients.

Molecular classification. We have shown, for most of the important cancer indications, that methylation patterns can be used to subclassify cancer in much the same way, and with much the same resolving power, as whole-genome mRNA expression microarrays [12]. In many instances, we can show that per data point (one CpG versus expression signal from one gene), methylation signals contain more highly compressed information than expression data. The reasons for this are unknown but may well be due to the "sharpness", reproducibility, and robustness of methylation techniques, as discussed above. Taken together with the advantages of a methylation testing platform as such, we see a great opportunity for methylation tests to complement or even replace expression microarrays in many of the applications that are most intensively discussed today. Possibly the clearest advantage of methylation technology in this application may be the ability to analyze very small amounts of dissected material from paraffinized samples that were previously analyzed by conventional pathology methods. This will allow seamless integration of a modern molecular methodologies with current clinical practices.

Pharmaco-epigenetic stratification. Similar generic technological arguments as for molecular classification can be used to argue that methylation will find an important place in pharmacogenetic testing. Several additional advantages can be named, too.

For pharmacogenetics and diagnostic classification of patients for clinical trials to work, genetic information (i. e., SNPs), as well as information on gene activity, is required. For example, the ability of an enzyme to metabolize a drug could be influenced by the protein sequence (i.e., a coding SNP) [20, 21], or its expression level (i.e., alteration of methylation), or both. And more often than not, information lying on an SNP could be compensated for by increased expression [4], leaving the SNP (or expression) information alone completely worthless in the individual case.

After more than 10 years of research, the absence of expression profiling methods in clinical research is conspicuous. The reason is the lack of robustness and reproducibility, difficulties in handling clinical mRNA samples by inexperienced clinicians, and the reluctance of pharmaceutical companies to develop a stratification method that cannot be plausibly turned into a diagnostic tool for routine testing of tens of thousands of patients. Methylation technology holds the promise of solving the technological problems that have so far hampered the introduction of gene-activity profiling into clinical development: First, it provides access to large numbers of archived specimens with well documented, comprehensive, clinical records. Retrospective analyses of well characterized specimen collections with detailed follow-up information are important in identifying markers for pathologic subclasses of cancer and even more so for markers for tumor aggressiveness and drug resistance, since long follow-up times may be necessary due to different disease courses. Second, once a diagnostic marker panel is identified, its applicability to routinely obtained, paraffin-embedded specimens greatly facilitates the next step: its introduction into large clinical trials and ultimately into clinical routine.

In genetics, problems occur not with the actual assay technology, but rather with isolating appropriate informative SNPs from small clinical populations. Current genetic methods do not work on the size of populations used even in phase III (let alone the preferred entry point at phase II) trials. Epigenomics integrates its methylation technology with the unique identification of linkage disequilibrium (ILD) technology to identify genetic markers for disease and drug response in clinical populations that are inaccessible to all other genetic technologies. ILD is derived from a method called "genomic mismatch scanning" (GMS), which isolates haplotypes that are identical by descent by producing heterohybrids of the genomes of two diseased individuals with a common founder. All genomic regions that contain mismatches at SNPs are then enzymatically degraded, allowing easy isolation of all haplotypes that contain putative disease loci [7].

The overall objective is to create strong predictive tests that are superior in terms of both genetics (SNPs) and epigenetics (gene activity through methylation). The combination of these technologies overcomes both critical disadvantages of other state-of-the-art technologies on one single platform, making genetics amenable to small clinical populations, and solves the assay-platform issues that have precluded the entry of mRMA microarray-based platforms into the routine testing market. Indeed, one of the conditions for incorporating test development for stratification purposes into end-point definitions of drug trials is the existence of an FDA-approved testing platform on which final marker panels are run during pivotal trials.

There is enormous price pressure on diagnostic tests today. Therefore, for clinical pharmacogenetics (and, eventually, personalized medicines) to gain a foothold in real-life pharmaceutical research and clinical diagnostics, both types of tests will have to be performed in an integrated fashion, and probably on one and the same technological platform, slashing platform investments for cost-constrained clinical institutions in half. Methylation analysis, reading expression information from DNA, is the only method that could possibly achieve this.

Monitoring. Monitoring patients for relapsing cancer relies on the same technological principles as the screening of asymptomatic people for early-stage disease, i.e., selective amplification from blood serum of fragments carrying a particular diagnostic methylation pattern against a background of normal, differently-methylated background fragments. For "simple" diagnosis of relapsing cancer, the same sets of methylation markers may be used as in screening for early-stage cancer.

However, for monitoring, additional questions are important and may be answered along with the diagnostic issues. For example, knowing whether a tumor is becoming chemoresistant could be of great value: the therapeutic regimen could be changed or adapted even before macroscopically visible metastases occur. One indicator of emerging drug resistance would be the overexpression – and concurrent hypomethylation – of multiple drug-resistance genes such as *MDR1* or other drug-resistance genes, a change in the hormone receptor status of cancer cells, or the appearance or disappearance of therapeutically useful surface antigens. Using methylation analyses, we expect that it will be possible to make such diagnoses in many cases.

Taken together, a comprehensive set of tests as described above, all of which could be based on methylation testing, has the potential to significantly change or even revolutionize the management of cancer.

9.2.3
Methylation, the Environment, and Lifestyle Diseases

A large body of evidence suggests that many diseases that are strongly modulated by environmental factors, as well as age-related diseases, are associated with, if not caused by, altered DNA methylation patterns in particular tissues. Aberrant DNA methylation in these diseases may be caused by three different mechanisms:

- Tissue-specific methylation patterns are established during cellular differentiation. In general, previously methylated genes become demethylated upon stimulation with transcription factors, thereby creating a repertoire of genes that are "ready to go" for later expression in this cell and, being a heritable signal, in successive cell generations. Indirect evidence suggests that environmental influences during this period may cause a failure to establish correct expression patterns in some genes, e.g., in hormone-responsive genes, through overstimulation or a lack of stimulation in the critical period. Although in theory reversible, these erroneous patterns may be difficult to erase and lead to predisposition of individuals to certain diseases later in life.
- Once established, tissue-specific methylation patterns are maintained quite stably throughout life. However, there is growing evidence that changes in the environ-

ment of cells or tissues leaves traces in the DNA methylation pattern which are likely to contribute to development of disease later in life. The traces of environmental influences may be targeted to specific genes, if specific metabolic pathways are stimulated or down-regulated with a subsequent change in their methylation patterns, or may occur randomly in the genome, e.g., through factors influencing the availability of methyl donor groups.

- A third mechanism may be the accumulation of methylation errors with age, which does happen [18–20], probably due to imperfect copying of methylation patterns to the daughter strands after replication by maintenance methyl transferase.

One archetypal disease which, albeit under the influence of genetic factors, is also influenced by the lifestyle of patients is type 2 diabetes mellitus (T2D). Indeed, type II diabetes is high correlated with obesity. Strict diet and exercise are excellent preventive measures against type II diabetes, even against the background of a genetic pre-disposition for obesity, further strengthening this notion. A range of circumstantial evidence suggests that all three proposed mechanisms contribute to the development of insulin resistance in type 2 diabetes mellitus:

- The level of prenatal glucose to which a person is exposed (e.g., in comparisons of siblings born before and after development of any type of diabetes in the mother) influences the likelihood of the child getting T2D [14]). This suggests the presence of a "cellular memory" in insulin target tissues such as adipose tissue, skeletal muscle, and liver.
- The promoters of several genes involved in glucose metabolism exhibit differential DNA methylation, e.g., the genes for facilitative glucose transporter 4 (GLUT4), the major glucose transporter in adipose and muscle tissues [15], and uncoupling protein 2 (UCP2) [16], a major candidate gene for the development of type 2 diabetes and an important regulator of energy expenditure. Establishment of correct methylation patterns of these genes may be disturbed in a critical period, or the patterns may change after environmental influences later in life.
- A general defect in DNA methylation in diabetes is suggested by the recent observation that S-adenosylmethionine (SAM), the main physiological donor of methyl groups, is decreased in the erythrocytes of diabetic patients. In addition, decreased erythrocyte concentrations of SAM and other alterations are associated with the disease progression [22].
- Recent insights into the pathogenesis of transient neonatal diabetes (TND), a rare subtype of diabetes that is characterized by transient hyperglycemia in the neonatal period and a predisposition to diabetes in adult life, provide a link between methylation, gene-dosage effects, and diabetes. Transient neonatal diabetes results from a doubling of the dosage of genes on chromosome 6q24. Paternal uniparental isodisomy, duplication of the respective band on 6q24, and loss of methylation in this imprinted region all result in phenotypically undistinguishable TND [17]. Interestingly, these individuals have an increased risk of developing type 2 diabetes later in life. Type 2 diabetes is age-related: not only is its incidence increased in older populations, but also the metabolic condition of individual patients dete-

riorates over time. DNA methylation errors that accumulate with increasing age could provide an explanation for both phenomena.

In a similar way, environmental influences and lifestyle factors may influence the development of arteriosclerosis. For example, nicotine influences the expression patterns of endothelial cells [42, 43]. Also, gene expression in aortic endothelial cells is influenced by flow conditions [44, 45]. Occurring over long periods of time, expression patterns tend to translate into altered methylation patterns, thereby propagating the effects even in the absence of further stimuli. A specific hypermethylation has been observed in proliferating aortic smooth-muscle cells compared with their normal counterparts [18].

Random DNA methylation changes occur in several tissue types during aging of organisms [18–20]. These accumulating age-related DNA methylation changes are involved in several diseases:

- In the colon, hypermethylation often starts in normal mucosa as a function of age and leads to field defects, with an increased risk of developing colorectal cancer ("acquired predisposition to colorectal neoplasia" [20]).
- Methylation-associated inactivation of the *ER*α gene in vascular tissue seems to occur specifically when the cells switch to a dedifferentiated, proliferating phenotype, but may also be associated with aging of the cardiovascular system [18].
- DNA methylation of the promoter region of the amyloid precursor protein gene, which is involved in the development of Alzheimer's disease, is reduced in with increasing age [21].

Taken together, the evidence suggests that methylation plays an important role in regulating gene expression, most likely including the expression of genes playing essential roles in the development of the metabolic syndromes and in so-called lifestyle-associated diseases. Different expression patterns that develop in association with changes in diet [23–25], in body weight [24, 25], and in exposure to environmental factors are likely to become "locked" by DNA methylation if they are present for a long period of time. DNA methylation, therefore, is likely to be involved in mediating the deleterious effects of increased body fat and high-fat diet on insulin sensitivity of insulin target tissues and on the cardiovascular system. Being a reversible modification, DNA methylation might also be involved in the adaptation of metabolism to starvation. On the other hand, metabolism of methyl groups may be affected by diet, body weight, and environmental factors [26], thus leading to untargeted, general hypomethylation of DNA in obese patients [22]. Moreover, DNA methylation errors have been shown to accumulate over time, contributing to many age-related diseases. These errors could add to the development of type 2 diabetes and cardiovascular diseases by reducing gene responsiveness (i.e., gene expression that needs to be adjusted to rapidly changing glucose and insulin levels).

9.2.4
Disease Gene Discovery

As outlined above, evidence is considerable that aberrant DNA methylation patterns are laid down early during cellular differentiation and may lead to the development of diseases such as type 2 diabetes and arteriosclerosis in later life. Also, DNA methylation patterns are subject to age-related changes, which may promote the development of metabolic and cardiovascular diseases with age. Also, environmental influences occurring later in life lead to gene-expression changes in exposed cells, which are most likely to be translated into more stable methylation patterns when they occur over a long period of time. The "locking" of environmental effects on the methylation patterns may predispose tissues to certain diseases such as T2DM and arteriosclerosis. Hence, for many diseases with a strong environmental component, methylation-pattern analysis constitutes a powerful tool for monitoring these changes and thereby identifying new potential targets for small molecules (in hypomethylation) or recombinant protein drugs (in hypermethylation) or identifying the involved pathways. If technical problems prevent the design of small-molecule drugs against an identified gene or protein or as a replacement for a down-regulated protein, knowledge of the involved pathways may open new possibilities for intervention. DNA-methylation profiling has several advantages compared to other genome-wide profiling approaches, e. g., mRNA expression analyses:

- DNA methylation picks up permanent expression changes rather than short-term alterations – fewer false positives occur (e. g., temporary expression changes).
- DNA methylation detection picks up changes that cannot be detected by any other method, since it exploits a different information layer.
- DNA methylation is suited for use in very large populations (e. g., from paraffin sections that are available through pathologies); therefore, subclasses can be detected – for example, the pathogenesis of type 2 diabetes is likely to be heterogeneous.
- DNA methylation has technical advantages, including the lack of particular requirements for sample handling (e. g., transfer between hospitals), and tissue samples can be analyzed retrospectively (e. g., paraffin-embedded tissues).

9.3
Tissue Engineering

9.3.1
The Development of Tissue Engineering

Tissue loss and end-stage organ failure pose major global health problems. Current attempts to overcome these problems mainly rely on transplantation of tissues or whole organs. Those approaches are limited by the availability of donor organs and their eventual immune rejection, situations unlikely to be ameliorated.

Tissue-engineering approaches are intended to provide treatments by implantation of engineered biological substitutes, alone or in combination with synthetic devices. Once this technology has been established, it should provide a sufficient supply of substitute tissues and should also circumvent problems of immune rejection (e. g., by using autologous tissues) [27].

Pioneering work in the tissue-engineering field was performed with skin replacement tissues as early as 1981. Engineered skin tissues contribute to superior treatment of severe wounds [30]. Also, cartilage replacement therapies have been introduced to the market and are used to treat worn-out cartilage [28]. In many other areas, such as heart-valve replacement [31], substitution of pancreatic beta cells in diabetes patients [33, 34], and replenishment of healthy dopaminergic neurons in Parkinson's patients [35, 36], tissue engineering technologies provide hope for new and better treatments of these currently incurable diseases.

Although the above technologies have a realistic possibility of being introduced within a reasonably short period of time, in the long range ex-vivo construction of whole organs is envisioned.

Irrespective of the particular therapeutic area researchers in tissue engineering are engaged in, they will come across at least one of the following steps in development:

- harvesting the original material (i. e., stem and other progenitor cells);
- maintaining or expanding cells without change in the phenotype or original differentiation status;
- manipulating and differentiating cells in a targeted, standardized, and efficient way to obtain the desired cell type;
- assessing the exact lineage, functionality, homogeneity, and differentiation status;
- proof of cost-effectiveness;
- post-surgical evaluation and follow-up quality-control monitoring.

Detailed analysis of the experimental outcome is required, regardless of the actual state of the research and differentiation experiments. In early steps of differentiation experiments, assessment of the results addresses questions as to whether correct progenitor cells were chosen or whether differentiation pathways are likely to yield correctly differentiated tissues. Correct differentiation and appropriate result assessment at these stages determine the chances of researchers to eventually come up with a fully developed, functional tissue. In more advanced stages of product development, assessment is required for the proof of product quality (QC) required for regulatory approval.

9.3.2
Complex Result Assessment in Tissue Engineering: an Unmet Need ...

Owing to the absence of adequate analytical methods, each of the above steps currently involves extensive use of poorly designed biochemical procedures. Assays are often inconclusive, lengthy, or costly; do not always qualify for prediction of the intended cellular function; and are meaningful only at the end of the differentiation process. Cell differentiation and effects of growth factors on differentiation can hardly be studied in detail. Also, arbitrary marker systems are used for tissue classification rather than a single, well understood, reproducible phenotyping system.

For the development of engineered tissues, this results in a situation in which:

- Efficient process optimization is difficult.
- Targeted identification of specific factors that promote growth or differentiation is extremely laborious.
- The lack of high-throughput capable methods for identifying factors inhibits the identification of chemical effector molecules.
- Product intermediates cannot be assessed efficiently as to whether they follow the correct differentiation pathways.
- Readily developed engineered tissues cannot be quality controlled in an unambiguous way.

As a consequence of these shortcomings, progress in research is sporadic, and only a few tissue-engineered cell products are now being marketed successfully. Industrial research, as well as funding for academic research, is limited, since the development itself is slow and expensive, relying on trial-and-error experiments [28]. Due to insufficient complexity of the currently applied methods, it is not possible to fully determine lineage and functionality of engineered cells and their congruence with the original cell type, leading to scientific ambiguity and increased public distrust.

Only complex analytical systems capable of determining selected – but principally genome-wide – changes can provide sufficient data for a thorough analysis of the cellular differentiation processes.

9.3.3
... That May Be Met by DNA-methylation Technologies

Epigenetic analysis can address the mentioned shortcomings, since it is an excellent surrogate method for analyzing differentiation pathways and for assuring the identity or bioactivity of differentiated cells and entire tissues.

DNA methylation not only modulates gene expression, it also functions as a cellular memory, which tells cells about the differentiation status that their progenitors had attained. It is assumed that persistent expression changes translate into the methylation pattern following the "loss-of-protection model", i.e., non-expressed genes become methylated. Methylation can be imagined as the "snap lock" that freezes differentiation states, which have been achieved through external stimuli, into the genome.

Upon bisulfite treatment, binary signals can be detected in homogenous tissues, which characterize a certain cellular state as a "0" or "1" signal (C is defined as 0 and ^mC is defined as 1) [29]. Hence, using several such signals, the differentiation stage of cells can be described as a digital string of "ones" and "zeros".

In contrast to conventional methods, DNA methylation reliably classifies cells so as to:

- provide a digital signal that is independent of the amount of original cellular material;
- provide principally genome-wide marker selection;
- analyze many CpG islands and genes simultaneously, obtaining complex patterns of results;
- allow the use of multiplexed markers;
- allow extensive sample logistics due to high stability of the DNA substrate;
- allow comparative analysis of in vitro tissue and in vivo follow-up after tissue transplantation;
- allow selective amplification of rare DNA identifying teratoma and other contaminant cells.

Taken together, determination of the DNA-methylation pattern opens the way for an iterative, reproducible, high-resolution analysis of cell status, which would meet the currently unmet need to select parameters characterizing cell differentiation. Consequently, specific tissue attributes could be identified and used as selection criteria for the quality of cell products, their differentiation status, and the functions of cells, tissues, and growth factors. With this tool, process optimization towards an efficient production of newly engineered tissues should be possible.

Also and in contrast with all other phenotyping systems, the methylation system meets requirements for a complex, robust, multi-parametrical testing tool, such as would be required by regulatory authorities. Thus, it could provide a gold-standard technology. As such, it could limit the current variability of markers tested to verify product quality. Researchers, industry, and regulatory agencies should welcome such an unambiguous, modular, and expandable standard technology serving as a cell-type classification system. A reliable quality standard is also desirable, since it provides planning security for companies desiring to develop tissue-engineering products.

If verified and certified by regulatory agencies, epigenetic quality control would not only address the current needs of the tissue-engineering community but may also stimulate a new wave of products, whose development is currently hampered by an unclear regulatory framework.

Taken together, tissue engineering may be aided by DNA-methylation technologies, which would allow for a more complex analysis of cells and:

- allow correct identification of cell types obtained from biopsies and autopsies;
- provide for development of methylation-based screening assays that ensure targeted development of tissue-engineered human cells;
- enable analysis of in vitro differentiated cells and thus correct identification of the cells;
- enable analysis of the cell-differentiation process and identification of growth factors, nutrient media, and other conditions that are required for correct differentiation;
- allow final quality-control assessment upon production of the tissues.

9.4
Methylation Therapy

9.4.1
Methylation Therapy

DNA-methylation patterns not only contain highly useful information regarding the differentiation state of a cell but can also be exploited to develop a new generation of highly sensitive, specific diagnostic tools. Moreover, epigenetic silencing of genes is a fundamental mechanism in tumorigenesis, therapeutic strategies reversing the silencing and leading to re-expression of these genes seem to hold great promise.

In the following sections, we describe several approaches that are currently under investigation in clinical or preclinical testing.

The DNA methylation machinery as a target. DNA methylation in promoter regions can be targeted by inhibiting DNA methyltransferases. The resulting reversal of hypermethylation in these regulatory regions leads to re-expression of genes that either suppress tumor growth or reestablish sensitivity to anticancer drugs [46].

The first DNA methyltransferase inhibitors that were synthesized are 5-aza-cytidine and its derivative 5-aza-2-deoxycytidine (decitabine) [47, 48]. As pyrimidine analogs, both agents are incorporated into genomic DNA and form a covalent complex with DNMT1, leading to its inactivation [49]. Since DNMT1 is mainly responsible for maintenance of DNA methylation, the newly synthesized DNA strands remain unmethylated. Decitabine has been tested in clinical trials for the treatment of several hematopoietic diseases, where it has differentiation-inducing effects [50]. Because of its toxicity, prolonged low-dose schedules [51] or combinations with other drugs [52] may prove superior to standard single-agent therapy. Decitabine restores sensitivity to other chemotherapeutic drugs in a xenograft model by re-inducing the expression of *hMLH1* [52], a mismatch-repair gene crucial for response to DNA-damaging drugs such as carboplatin and epirubicin. Hence, strong synergies with conventional cytotoxic drugs can be expected, and decitabine may prove highly valuable in dealing with a priori or acquired drug resistance of many cancers.

A different approach to inhibiting the DNA-methylation machinery has been taken by Methylgene, a company that is currently running phase I and II clinical trials of MG98, an antisense oligonucleotide against DNMT1. Treatment of human tumors with MG98 reduces the levels of DNMT1 protein and induces re-expression of several genes that were silenced in the tumors, including the *p16* tumor-suppressor gene [53].

The major concern with nonspecific gene demethylation is that global demethylation may lead to reactivation of silenced proviral sequences or imprinted genes with tumor growth-promoting effects. Inhibition of methyltransferases that are mainly responsible for de novo methylation, such as DNMT3A and DNMT3B, can be expected to not be associated with a danger of demethylating these elements, but may lack the strong demethylating effect of DNMT1 inhibition.

The chromatin as a target.   Another highly promising target for epigenetic therapy is histone modification. Histone modifications influence gene expression by remodeling the chromatin into active or inactive states [54]. DNA methylation and histone modifications affect each other in a variety of ways [55–57]. The histone modification that has received the most attention regarding its systematic manipulation is deacetylation. Deacetylation of lysine residues, predominantly in histones H3 and H4, is associated with silencing of genes. Several families of histone deacetylases (HDACs) and histone acetyl transferases (HATs) have been identified. For therapeutic purposes, HDACs can be inhibited by several classes of compounds: short-chain fatty acids (e.g., butyrate), hydroxamic acids (e.g., trichostatin A, suberoylanilide hydroxamic acid (SAHA), oxamflatin), cyclic tetrapeptides (e.g., trapoxin A, apicidin), and benzamides (e.g., MS-275) [58]. In human tumor cells, HDAC inhibitors induce growth arrest, cellular differentiation, and apoptosis [55, 58]. Interestingly, only about 2% of the genes changed their expression levels after treatment with HDAC inhibitors, among them the tumor suppressor and cell cycle regulator *p21* [58, 59]. As might have been anticipated, due to the intensive interplay among HDACs and the DNA-methylation machinery, a combination of inhibition of DNA methyltransferases and of HDAC yields synergistic effects on the re-expression of silenced genes [60]. Following these highly promising preclinical studies, several HDAC inhibitors are currently undergoing early phases of clinical trials (SAHA, depsipeptide, MS-27–275).

Crystallographic studies have revealed the structure of the catalytic site of HDACs, and competition with substrates for the catalytic center has been identified as the mechanism of action of hydroxamic acid HDAC inhibitors such as TSA and SAHA [60, 61]. These findings should greatly enhance the development of more specific HDAC inhibitors. This is of particular importance, since most naturally occurring HDAC inhibitors are thought to have additional effects besides their HDAC inhibition – probably as a consequence of evolutionary selection for their biological role in killing enemies and competitors.

Again, as with unspecific demethylating agents, the question arises of how to target the effects to specific genes, the re-expression of which are needed to restore apoptotic or tumor growth-suppressing pathways or sensitivity to other anticancer agents.

Targeting specific sites: the Sangamo approach.   Sangamo Biosciences has undertaken a promising approach that may well overcome the problem of unspecific activation of genes by both DNA demethylating agents and HDAC inhibitors. To direct a specific activity (acetyl transferase, histone deacetylase, K9 methyl transferase, etc.) to a particular site in the genome, the most common DNA binding motif found in nature, the Cys2-His2 zinc-finger DNA binding domain, is used as a scaffold for constructing factors with the desired specificity for a particular DNA binding site. Depending on whether re-expression of a gene or its down-regulation is the therapeutic goal, a functional domain that acts as a transcriptional activator or repressor can be coupled to the DNA-binding domain of the artificial peptide. Among the most interesting functional domains are factors targeting the chromatin surrounding the transcription-factor binding site, e.g., domains with HDAC or HAT activity or domains influencing K9 methylation. Due to its gene specificity, this approach is expected to control gene function much more efficiently than unspecific approaches and can be presumed to be associated with fewer side effects [62].

9.5
Agriculture

For centuries, animal breeders have improved economically important traits (e.g., productivity, size, weight, etc.) in farm animals by selecting superior animals as progenitors for the next generation. Today, desirable characteristics have shifted towards traits such as disease resistance, feeding efficiency, and longevity, since they appear to harmonize claims for economically and ecologically responsible farming. Corresponding traits should therefore be a focus of animal breeding programs. For many of these traits, the efficiency of traditional quantitative genetic approaches is limited, due to late onset of phenotypic perceptibility or difficulty with recording the phenotypes. To pursue genetic improvement of these characteristics, animal breeders need to take advantage of molecular analysis of complex quantitative traits.

When originally introduced to animal breeding, modern genetics was mainly limited to genetic traits determined by single genes. In these traits, molecular diagnostics could be developed to detect affected alleles in animals used for breeding. Genetic traits of this monogenic class show a Mendelian inheritance pattern. Animals that test positive for certain traits can be excluded from the active breeding population to prevent the spread of undesired traits or included to promote the spread of desired traits. Unfortunately, monogenic traits constitute a minority of the useful characteristics.

Current economically interesting phenotypes are mostly low-heritable complex traits, many of which are influenced by environmental effects. Complex traits (also referred to as multifactorial traits) are controlled by several genes and subject to environmental influences. The genes for these traits have quantitative (additive) effects on the phenotype that are not subject to classical Mendelian heredity. These multifactorial traits are characterized by continuous phenotypes, leading to the expression "quantitative trait loci" (QTL).

Use of molecular QTL information should improve selection intensity and accuracy. It would allow breeders to select breeding stocks based on molecular genetic information.

Since the economic value of a marker set depends on the percentage of the molecular variance explained, a maximal amount of information should be captured by using different methods and taking into consideration as many information levels as possible. At the level of gene regulation, two equally important sets of regulatory mechanisms can play a role:

- genetic variability in control regions of genes (e.g., by altering the binding sequences for transcription factors);
- epigenetic modification in control regions of genes (e.g., changing the methylation patterns of a CpG position in a transcription factor binding sequence).

Elucidation of both mechanisms appears to be needed for efficient analysis of the causes of multifactorial, economically important phenotypes.

The principal technical approach taken to identify relevant genetic influences and inheritance patterns is the identification of linkages based on marker maps. This approach is aided by the "candidate gene approach", which is used to narrow the selection of genes that may be involved in certain traits by functional correlation. Despite constant technical improvements and the design of new technologies [37], the analysis remains costly and inefficient. Also, it cannot identify those components of traits that are under epigenetic control.

The archetypal epigenetic regulation mechanism is DNA methylation, analysis of which may therefore significantly contribute to the identification of QTLs. Interindividual variability of methylation patterns can be generated by various biological influences. Thus, it appears reasonable to assume that methylation-dependent gene activity comprises the whole phenotypic continuum of a QTL, i.e., a certain trait can be influenced significantly by increased or decreased methylation in the regulatory region of one of several genes. Recent technical progress has resulted in the ability to perform high-throughput, whole-genome methylation studies.

Genomic imprinting is another methylation-dependent phenomenon with potential significance in QTL research. The role of imprinting in determining body composition, which is closely related to the economically important parameter of meat quality, was investigated by de Koning [38]. The authors studied an experimental cross between Chinese Meishan pigs and commercial Dutch pigs. A whole-genome scan revealed evidence for five quantitative-trait loci affecting body composition, of which four were imprinted. Imprinting was tested with a statistical model that separated the expression of paternally and maternally inherited alleles. For backfat thickness, a paternally expressed QTL was found on SSC2. In the same region of SSC2, a maternally expressed QTL affecting muscle depth was found. Chromosome 6 harbored a maternally expressed QTL on the short arm and a paternally expressed QTL on the long arm, both affecting intramuscular fat content. The individual QTL explained 2%–10% of the phenotypic variance. Using a similar approach, Nezer et al. [63] mapped an imprinted QTL having major effects on muscle mass and fat deposition to the *IGF2* locus of pigs.

Additional use can made of the fact that a QTL is controlled by imprinting: conventional techniques often produce a very imprecise gene location (within 20–40 cM). Methylation analysis might effectively accelerate the gene-finding process within such a large genomic area, because one could concentrate the gene search on the differentially methylated region.

The basis of a QTL producing epigenetic variation could be a disturbed balance of genomic imprinting between the parental chromosomes or a modification of an imprinting center (IC). The variation could be recognized by regional analysis of the methylation pattern. That means that methylation may become important for the molecular characterization of economically attractive traits in farm animals on two different levels: the study of methylation patterns could reveal both interindividual variability in gene expression and imbalance of the imprinting situation between parental chromosomes or genes. These molecular phenomena would significantly contribute to phenotype variation and might become the basis for relevant QTL-marker tests for use in animal breeding programs.

9.6
Outlook

In this chapter we discussed applications of methylation sciences that are so far in the future that no products have yet reached the market. However, millions of dollars of venture capital have been invested in all the areas treated here and are even now being used in development of mature products.

The potential of methylation science can be glimpsed if we recall that these few pages go through a list of many of the major sectors of the life sciences, diagnostics, pharmaceutical development, personalized medicine, and agriculture, in each of which we find highly attractive opportunities. At the same time, we discussed several completely different disease indications: we discussed disease management and diagnostics exclusively for oncology, and we illustrated uses in pharmaceutical research and development mainly for metabolic diseases. The tremendous potential of methylation as an applied science becomes obvious if we believe that the diagnostic applications expand into metabolic, cardiovascular, and other major human diseases. Likewise, pharmaceutical research and development can be supported with methylation information in as many disease indications. We believe that this is so, indeed attributing to methylation an importance that could be comparable to that of the other "big" genome sciences. We have argued that methylation is the only parameter that truly changes genome function in aging, as well as being affected by environmental influences. Therefore, we expect methylation to be not only competitive as a tool for diagnostics and research, but in many cases but simply irreplaceable. All in all, we expect methylation technologies to become a firm component of most, if not all, genomics-based research and development in the future.

References

1 CHAN, M. F.; LIANG, G.; JONES, P. A. Relationship between transcription and DNA methylation, *Curr. Top. Microbiol. Immunol.* **2000**, *249*, 75–86.

2 CHRISTMANN, M. et al. Acquired resistance of melanoma cells to the antineoplastic agent fotemustine is caused by reactivation of the DNA repair gene MGMT, *Int J. Cancer,* **2001**, *92*, 123–129.

3 ESTELLER, M. et al. Inactivation of the DNA-repair gene MGMT and the clinical response of gliomas to alkylating agents, *New Engl. J. Med.* **2000**, *343*, 1350–1354.

4 HAMMONS, G. J. et al. Specific site methylation in the 5′-flanking region of CYP1A2 interindividual differences in human livers, *Life Sci.* **2001**, *69*, 839–845.

5 OSTENSON, C. G. The pathophysiology of type 2 diabetes mellitus: an overview. *Acta Physiol. Scand.* **2001**, *171*, 241–247.

6 SMILINICH, N.J.; DAY, C. D. et al. A maternally methylated CpG island in KvLQT1 is associated with an antisense paternal transcript and loss of imprinting in Beckwith-Wiedemann syndrome, *Proc. Natl. Acad. Sci. USA* **1999**, *96*, 8064–8069.

7 NELSON, S. F. et al. Genomic mismatch scanning: a new approach to genetic linkage mapping, *Nature Genetics* **1993**, *4*, 11–18.

8 GRANT, D. J.; SHI, H.; TENG, C. T. Tissue and site-specific methylation correlates with expression of the mouse lactoferrin gene, *J. Mol. Endocrinol.* **1999**, *23*, 45–55.

9 CONDORELLI, D. F. et al. A neural-specific hypomethylated domain in the 5′ flanking region of the glial fibrillary acidic protein gene, *Dev. Neurosci.* **1997**, *19*, 446–456.

10 DREOSTI, I. E. Nutrition, cancer, and aging, *Ann. N. Y. Acad. Sci.* **1998**, *854*, 371–377.

11 COFFIN, J. C. et al. Effect of trihalomethanes on cell proliferation and DNA methylation in female B6C3F1 mouse liver, *Toxicol. Sci.* **2000**, *58*, 243–252.

12 ADORJAN, P. et al. Tumor class prediction and discovery by microarray-based DNA-methylation analysis, *Nucleic Acids Research* **2002**, *30*, e21.

13 MODEL, F. et al. (2001) Feature selection for DNA methylation based cancer classification. Bioinformatics 17 Suppl 1:S157–64.

14 DABELEA, D. et al. Intrauterine exposure to diabetes conveys risks for type 2 diabetes and obesity, *Diabetes* **2000**, *49*, 2208–2211.

15 YOKOMORI, N. et al. DNA demethylation during the differentiation of 3T3-L1 cells affects the expression of the mouse GLUT4 gene, *Diabetes* **1999**, *48*, 685–690.

16 CARRETERO, M. V. et al. Transformed but not normal hepatocytes express UCP2, *FEBS Lett.* **1998**, *439*, 55–58.

17 TEMPLE, I. K. et al. Transient neonatal diabetes: widening the understanding of the etiopathogenesis of diabetes, *Diabetes* **2000**, *49*, 1359–1366.

18 POST, W. S. et al. Methylation of the estrogen receptor gene is associated with aging and atherosclerosis in the cardiovascular system, *Cardiovasc. Res.* **1999**, *43*, 985–991.

19 BORNMAN, D. M. et al. Methylation of the E-cadherin gene in bladder neoplasia and in normal urothelial epithelium from elderly individuals, *Amer. J. Pathol.* **2001**, *159*, 831–835.

20 ISSA, J. P. et al. Accelerated age-related CpG island methylation in ulcerative colitis, *Cancer Res.* **2001**, *61*, 3573–3577.

21 TOHGI, H. et al. Reduction with age in methylcytosine in the promoter region −224 approximately −101 of the amyloid precursor protein gene in autopsy human cortex, *Brain Res. Mol. Brain Res.* **1999**, *70*, 288–292.

22 POIRIER, L. A. et al. Blood S-adenosylmethionine concentrations and lymphocyte methylenetetrahydrofolate reductase activity in diabetes mellitus and diabetic nephropathy, *Metabolism* **2001**, *50*, 1014–1018.

23 ANDREELLI, F. et al. Regulation of gene expression during severe caloric restriction: lack of induction of p85 alpha phosphatidylinositol 3-kinase mRNA in skeletal muscle of patients with type II (non-insulin-dependent) diabetes mellitus, *Diabetologia* **2000**, *43*, 356–363.

24 NADLER, T. S. et al. The expression of adipogenic genes is decreased in obesity and diabetes mellitus, *Proc. Natl. Acad. Sci. USA* **2000**, *97*, 11371–11376.

25 DUCLUZEAU, P. H. et al. Regulation by insulin of gene expression in human skeletal muscle and adipose tissue: evidence for specific defects in type 2 diabetes, *Diabetes* **2001**, *50*, 1134–1142.

26 POIRIER, L. A. et al. Blood determinations of S-adenosylmethionine, S-adenosylhomocysteine, and homocysteine: correlations with diet. *Cancer Epidemiol. Biomarkers Prev.* **2001**, *10*, 649–655.

27 VACANTI, J.; VACANTI, C. (**2001**) The history and scope of tissue engineering. In *Principles of Tissue Engineering*, Lanza, R.; Langer, R.; Vacanti, J. eds., Academic Press, San Diego, 1–7.

28 BROMLEY, A. (**2001**) *Tissue Engineering; Technologies and Markets*, Clinica Reports, CBS 883. PJB Publications.

29 FROMMER, M. et al. A genomic sequencing protocol that yields a positive display of 5-methylcytosine residues in individual DNA strands, *Proc. Natl. Acad. Sci. USA* **1992**, *89*, 1827–1831.

30 BELL, E. et al. Living tissue formed in vitro and accepted as skin-equivalent tissue of full thickness. *Science* **1981**, *211*, 1052–1054.

31 LOVE, J. W. (**2001**) Cardiac prosthesis. In *Principles of Tissue Engineering*, Lanza, R.; Langer, R.; Vacanti, J. eds., Academic Press, San Diego, 455–467.

32 GERHART T. N. et al. Healing segmental femoral defects in sheep using recombinant human bone morphogenetic protein, *Clin Orthop.* **1993**, *293*, 317–326.

33 SHAPIRO, A. M. et al. Islet transplantation in seven patients with type 1 diabetes mellitus using a glucocorticoid-free immunosuppressive regimen, *New Engl. J. Med.* **2000**, *343*, 230–238.

34 RAMIYA, V. K. et al. Reversal of insulin-dependent diabetes using islets generated in vitro from pancreatic stem cells, *Nat. Med.* **2000**, *6*, 278–282.

35 LINDVALL, O. et al. Grafts of fetal dopamine neurons survive and improve motor function in Parkinson's disease, *Science* **1990**, *247*, 574–577.

36 LINDVALL, O. Update of fetal transplantation: the Swedish experience, *Movement Disord.* (**1998**), *13* (Suppl.1), 83–87.

37 CHEUNG, G. et al. Linkage-disequilibrium mapping without genotyping, *Nature Genetics* **1998**, *18*, 225–230.

38 DE KONING, D. J. et al. Genome-wide scan for body composition in pigs reveals important role of imprinting, *Proc. Natl. Acad. Sci. USA* **2000**, *97*, 7947–7950.

39 LEFEBVRE, L. et al. Abnormal maternal behaviour and growth retardation associated with loss of imprinted gene *Mest*. *Nature genetics* **1998**, *20*, 163–169

40 WU, T. L. et al. Cell-free DNA: measurement in various carcinomas and establishment of normal reference range, *Clin. Chim. Acta* **2002**, *321*, 77–87.

41 SOZZI, G. et al. Analysis of circulating tumor DNA in plasma at diagnosis and during follow-up of lung cancer patients, *Cancer Res.* **2001**, *61*, 4675–4678.

42 ZHANG, S. et al. Nicotine induced changes in gene expression by human coronary artery endothelial cells, *Atherosclerosis* **2001**, *154*, 277–283.

43 ZHANG, S. et al. Microarray analysis of nicotine induced changes in gene expression in endothelial cells, *Physiol. Genomics* **2001**, *5*, 187–192

44 BROOKS, A.R. et al. Gene expression profiling of human aortic endothelial cells exposed to disturbed flow and steady laminar flow, *Physiol. Genomics* **2002**, *9*, 27–41.

45 GARCIA-CARDENA, G. et al. Mechanosensitive endothelial gene expression profiles: scripts for the role of hemodynamicx in atherogenesis? *Ann. N.Y. Acad. Sci.* **2001**, *947*, 1–6.

46 STRATHDEE, G.; BROWN, R. Aberrant DNA methylation in cancer: potential clinical interventions, *Expert Rev. Mol. Med.* **2002**.

47 PISKALA, A.; SORM, F. Nucleic acids components and their analogues: synthesis of l-glycosyl derivatives of 5-azauracil and 5-azacytosine. *Coll. Czeck Chem. Commun.* **1964**, *29*, 2060–2076.

48 PLIML, J.; SORM, F. Synthesis of 2-deoxy-D-ribofuranosyl-5-azacytisine. *Coll. Czeck Chem. Commun.* **1964**, *29*, 2576–2577.

49 BOUCHARD, J.; MOMPARLER, R. L. Incorporation of 5-aza-2-deoxycytidine and 5'-triphosphate into DNA: interactions with mammalian DNA polymerase and DNA methylase. *Mol. Pharmacol.* **1983**, *24*, 109–114.

50 PINTO, A.; ZAGONEL, V. 5-aza-2'-deoxycytidine (decitabine) and 5-azacytidine in the treatment of acute myeloid leukemias and myelodysplastic syndromes–past, present and future trends, *Leukemia* **1993**, *7*, 51–60.

51 LUEBBERT, M. et al. Cytogenetic responses in high-risk myelodysplastic syndrome following low-dose treatment with the DNA methylation inhibitor 5-aza-2'-deoxycytidine. *Br. J. Haematol.* **2001**, *114*, 349–357.

52 PLUMB, J. A. et al. Reversals of drug resistance in human tumour xenografts by 2'-deoxy-5-azacytidine-indueced demethylation of the hMLH1 gene promoter, *Cancer Res.* **2000**, *60*, 6039–6044.

53 REID, G. K. et al. Selective inhibition of DNA methyltransferase enzymes as a novel strategy for cancer treatment, *Opin. Mol. Ther.* **2002**, *4*, 130–137.

54 TYLER, J. K.; KADONAGA, J. T. The dark side of chromatin remodelling: repressive effects on transcription, *Cell* **1999**, *99*, 443–446.

55 Cervoni, N.; Szyf, M. Demethylase activity is directed by histone acetylation, *J. Biol. Chem.* **2001**, *276*, 40778–40787.

56 Fuks, F. et al. DNA methyltransferase Dnmt1 associates with histone deacetylase activity. *Nat. Genet.* **2001**, *24*, 88–91.

57 Robertson, K. D. et al. DNMT1 forms a complex with Rb, E2F1 and HDAC1 and represses transcription from E2F-responsive promoters, *Nat. Genet.* **2001**, *25*, 338–342.

58 Marks, P. A. et al. Histone deacetylase inhibitors as new cancer drugs, *Curr. Opin. Oncol.* **2001**, *13*, 477–483.

59 Van Lint, C. et al. The expression of a small fraction of cellular genes is chan-

ged in response to histone hyperacetylation, *Gene. Expr.* **1996**, *5*, 245–253.

60 Cameron, E. E. et al. Synergy of demethylation and histone deacetylase inhibition in the re-expression of genes silenced in cancer, *Nature Genet.* **1999**, *21*, 103–107.

61 Finnin, M. S. et al. Structures of a histone deacetylase homologue bound to the TSA and SAHA inhibitors, *Nature* **1999**, *401*, 188–193

62 Reik, A. et al. Biotechnologies and therapeutics: chromatin as a target, *Curr. Opin. Genet. Dev.* **2002**, *12*, 233–242.

63 Nezer, C. et al. An imprinted QTL with major effect on muscle mass and fat deposition maps to the IGF2 locus in pigs, *Nature Genetics* **1999**, *21*, 155–156.

Subject Index